Springer Japan KK

Yoshitaka Okada

Competitive-*cum*-Cooperative Interfirm Relations and Dynamics in the Japanese Semiconductor Industry

Springer

OKADA Yoshitaka
Professor of Sociology and Business
Faculty of Comparative Culture, Sophia University
4 Yonban-cho, Chiyoda-ku, Tokyo 102-0081, Japan

ISBN 978-4-431-67984-4

Library of Congress Cataloging-in-Publication Data

Okada, Yoshitaka, 1949-
    Competitive-cum-cooperative interfirm relations and dynamics in
the Japanese semiconductor industry / Yoshitaka Okada.
      p.   cm.
    Includes bibliographical references and index.
    ISBN 978-4-431-67984-4   ISBN 978-4-431-67923-3 (eBook)
    DOI 10.1007/978-4-431-67923-3
    1. Semiconductor industry—Japan. 2. Integrated circuits
industry—Japan. 3. Strategic alliances (Business)—Japan.
4. Interorganizational relations—Japan. 5. Competition—Japan.
I. Title.
HD9696.S43J354 2000
338.4'762138152'0952—dc21   99-41456

Printed on acid-free paper

© Springer Japan 2000
Originally published by Springer-Verlag Tokyo in 2000
Softcover reprint of the hardcover 1st edition 2000

SPIN: 10736360

Table of Contents

Chapter Two
Technological Development, Corporate Strategies and Market Competition   53

Chapter Three
Basic Characteristics of Intra- and Interfirm
Interaction   85

Chapter Four
Effects of Cooperation-oriented CCC Interaction
in Procurement Relations   113

Chapter Five
Effects of Competition-oriented CCC Interaction in Sales and Internalized Interaction in R&D   155

Chapter Six
CCC Dynamics and Structure of Intra- and Interfirm Relations   179

List of Tables

List of Figures

Preface

This book is the cumulative outcome of several projects. My interest in the semiconductor industry started when I was involved in a United Nations University project titled "Japan's Self-reliance in Science and Technology for National Development." The late Professor Toshio Shishido, then the Vice President of the International University of Japan, generously included me in the project as a young scholar along with prominent Japanese specialists in technology and development.

My interest in institutional analysis began when I was invited to be visiting assistant professor at the University of Wisconsin-Madison in 1987. I am grateful to Professor Solomon Levine for providing me with such a rare opportunity. There I met Professor Rogers Hollingsworth, who introduced me to the fascinating field of institutional analysis and to a group of top-level scholars working in the field. Inspired by institutional analysis, I began to develop the conceptual framework needed to identify the institutional dynamics of Japanese high-technology industries. Grants from the International University of Japan (1989) and the Japanese Ministry of Education (1990-1991) enabled my colleagues and me to start the project. My role was to investigate the semiconductor industry; Professors Shinichi Watanabe and Ichiro Inukai of the International University of Japan were to inquire into bio-related industries. They gave me invaluable assistance by teaching me an economist's perspective and enriching my conceptual framework. Later, small grants from the Kajima Foundation (1992) and IBM Japan (1993) enabled me to continue research on the semiconductor industry. Travel grants from the International University of Japan (1993), the Nomura International Foundation (1994), the Murata Science Foundation (1996), and the Japan Foundation (1996) have enabled me to pursue and present this research.

In conducting a survey and interviews, I am deeply indebted to the late Professor Saburo Okita, the former Foreign Minister and President of the International University of Japan, for his human dignity and his sincere understanding of young scholars. He opened many doors to facilitate interviews with prominent businessmen and to conduct the company surveys. Also greatly appreciated is the assistance offered by several top-level managers of Japanese semiconductor companies for allowing me to conduct research. Needless to say, I am especially thankful to the twenty-nine managers in the seven semiconductor companies who cooperated with me. Without their patience and expertise, this research would not have been possible.

My conceptualization of competitive-cum-cooperative governance has been advanced through the "Comparison of Capitalist Economies" project organized by Professor Rogers Hollingsworth. My work has benefited from criticism by Professor Richard Whitley of the University of Manchester Business School, who commented on my paper in the capitalist-economies project.

The theory of competitive-cum-cooperative interfirm governance presented in this book is the outcome of the intellectual stimulation that I received from friends, company managers, and many scholars with whom I interacted. Ms. Ellen Jane Hollingsworth, the former Research Director of the Mental Health Research Center, the University of Wisconsin-Madison, and Professor Richard Whitley have given me valuable comments on this manuscript. Fr. Robert Ballon and Ms. Yasuko Hamabata of Sophia University have consistently provided me with valuable assistance when I most needed it: during the Advanced Development Management Program and the last stage of editing this book. I am grateful to Ms. Lisa Schreibersdorf for correcting my English and Mr. David Gear and Mr.Carlos Roberto Salas for assisting me at several stages of completion of this manuscript. Of course, none of them is responsible for the content or for any errors.

I am grateful to the Ministry of Education for providing me with a Grant-in-Aid for Publication of Scientific Research Results and to the Japan Society for the Promotion of Science for administering the grant.

My deepest appreciation goes to my family members. My wife, Hideko, and my children, Mika and Erina, have been silently and patiently supporting this book project. They did not complain about many days and long hours of work, even when my work spoiled joyful family plans. Our dog, JJ, often lessened my fatigue by showing her joy and by sleeping at my feet during many late nights.

Yoshitaka OKADA
Madison, Wisconsin
September 1999

Introduction

In the era between World War II and the early 1990s, a drastic transformation in the world's division of industrial labor took place. The recovery of West German industries, the rise of Japanese industrial capabilities, the rapid catch-up of newly industrialized economies (NIEs), the high growth rates of Southeast Asian countries and China, and the relative decline of U.S. international competitiveness changed the world's industrial power configuration. The distribution of industrial power in the world became increasingly multi-polar, which created unprecedented international competition among developed countries and NIEs. Developed countries, facing constant threats from other developed countries and from rapidly approaching NIEs, searched for avenues to strengthen or maintain their international competitiveness by developing highly sophisticated technologies and by swiftly transforming their domestic industrial structure. Questions were constantly raised by government officials, businessmen, and scholars as to how international competitiveness in high technology industries could be maintained or enhanced. Their attention was not limited only to types of management or technology, but was extended to include institutional arrangements.

Japan has often been the focus of this inquiry, since it has succeeded in enhancing its industrial capability at a very rapid pace. Its nominal GDP grew from US$15.7 billion in 1953 to US$4,190 billion in 1993, and its nominal GDP per capita increased from US$181.1 to US$33,701 (Kokusai Rengo, 1963:493; United Nations, 1979:137; Keizai Koho Senta, 1995:11). And since the middle of the 1970s, it has shifted its efforts toward enhancing high technology industries.

The semiconductor industry is a good example of this successful transformation. Despite U.S. leadership in semiconductor technologies in the 1950s and 1960s, Japanese success in developing a very large-scale integrated circuit (VLSI) in 1977 enabled Japan to dominate the area of dynamic-random-access memory (DRAM) integrated circuits (ICs) and to gradually

1

strengthen its competitiveness in other areas of semiconductors (Okada, 1989a, 1989b, 1990).

In the 1980s, Japan came to dominate the 64K (70% in 1982), 256K (90% in 1984), and 1M (90% in 1988) bit DRAM world markets (Denpa Shinbunsha, 1983-1989; Press Journal, 1985). As the number of DRAM makers in the U.S. declined from fourteen in 1970 to three in 1986, the Japanese world market share of all types of semiconductors first came close to the U.S. in 1985, then grew even bigger in 1987 (Japan=48% vs. U.S.=39%), and peaked in 1988 (Japan=about 51% vs. U.S.=about 37%) (Department of Defense, 1987; Denpa Shinbunsha, 1988; Nihon Denshi Kikai Kogyo Kai, 1994:3). Among the top ten semiconductor producers in terms of sales, Japanese companies dominated the top three positions (NEC, Toshiba, and Hitachi) and three other positions in 1988 (Denpa Shinbunsha, 1989). Such success not only generated trade disputes between the U.S. and Japan, but also came to influence some U.S. innovation systems.[1]

The Japanese world market share, however, declined a little below the U.S. with the revitalization of the U.S. semiconductor industry and the advancement of Korean companies in 1993 (Nihon Denshi Kikai Kogyo Kai, 1994:3). The top semiconductor sales position went to a U.S. manufacturer, INTEL, while a Japanese company, NEC, retained the second position (Denpa Shinbunsha, 1995:843). For the first time, a Korean company, Samsung, came to lead the sales of 16M bit DRAM in 1993 (Press Journal, 1994:164) and occupied the seventh position in the sales of semiconductors (Denpa Shinbunsha, 1995:842). Although the golden age of the Japanese semiconductor industry (1977-1992) seems to be over, Japanese companies still maintain technological leadership in DRAM and diverse other semiconductor areas.[2]

---

[1] Learning from the Japanese experience, the U.S. Government recognized the importance of cooperative research among private companies, previously considered a violation of the Anti-Trust Law. The National Cooperative Research Act passed in 1984 legalized cooperative research, and allowed the government and the private sector to create cooperative ventures. For example, the Defense Department and the private sector jointly established SEMATEC, while a number of companies cooperatively established the Microelectronics and Computer Technology Corporation (MCC) and the Semiconductor Research Corporation (SRC).

[2] Finan and Frey (1994) term the period between 1960 and 1984 the golden age of the Japanese electronics industry and identify 1984-1991 as a declining period. The Japanese semiconductor industry shows a delayed cycle. The golden age started when the VLSI cooperative succeeded in developing 64K-bit DRAM in 1977 and continued as long as Japanese companies kept leadership in the world market share. It ended in 1992 when they came to share the top position in the world market with U.S. companies.

How did the Japanese semiconductor industry achieve its dynamics? Many management theories attribute the success of Japanese dynamics to measures that promote cooperation and trust in intra- and interfirm relations (Gerlach, 1989; Inoue, 1985; Stowsky, 1989; Nakatani, 1984; Uekusa, 1987; Kinzley, 1991; Iwata, 1977, 1982, 1984). Is cooperation and trust sufficient to explain such dynamics? To make cooperative relations dynamic, isn't it necessary to make some use of stringent measures, such as introducing competition among cooperating members?

The Japanese achievement in the DRAM market was largely indebted to the success of a cooperative research association for developing processing technologies necessary for very large scale integration (the VLSI Cooperative). With financial sponsorship, the Japanese Government succeeded in obtaining cooperation among big semiconductor and semiconductor-manufacturing-equipment producers.[3] Undeniably, interfirm cooperation among competing companies and between companies of complementary specialization was one of the crucial factors that allowed Japanese companies to take the world leadership in the DRAM area.

Japanese companies cooperate with others over management, finance, technology, production and R&D. Such practices are well known and include the following complex programs and processes: cooperative quality assurance programs; value analyses and engineering to reduce costs by analyzing many different facets of production and process technology, products, inventory, marketing, etc.; plans and strategies for reducing prices (strategic pricing) that are jointly developed among long-term partners; and just-in-time production (JIT). R&D is often jointly conducted to maximize gains from the know-how of partners; sometimes organizations even make use of the same computers in designing and drawing (simultaneous engineering). In order to develop the capability of small- and medium-sized partners, large-sized companies often provide necessary inputs, management and technical training, and financial assistance.

Even in the sales activity of semiconductor companies, a key to winning in severe market competition is shifting from simple market-oriented activities, to becoming a cooperative partner on projects by suggesting diverse ideas and sharing information. The former approach was effective in the 1970s, but these days it often leads to a losing battle. Finding suppliers is determined in much earlier stages. In 1992 about 50% of a sales person's

---

[3] Cooperation is understood as joint or collaborative behavior directed toward some goal based on common interest and mutual expectations.

visits were, in an estimate, to the product development and engineering sections of a company, and the proportion has increased significantly since then to about 70% in the middle of the 1990s. Such competition in the early stages of product development requires developing a strong sense of cooperation with the personnel in charge of planning, product development, designing, and engineering.

There is abundant evidence that Japanese interfirm relations are a very important source of generating dynamic development in the semiconductor industry. And cooperative behavior is becoming a very important factor in deciding who wins in the marketplace. It is, however, difficult to understand how cooperation works as the key factor for explaining interfirm dynamics. What is often hidden within cooperative tactics is competition generated by cooperative activities or competitive measures that help cooperation to be dynamic. Without competition, cooperation itself might not contribute to interfirm dynamics.

Within the VLSI cooperative, what is known as parallel research was taking place in each company. At the same time that researchers from competing semiconductor companies cooperated under the joint laboratory of the VLSI cooperative, each company established a special research team within its own laboratory, and duplicated all research activities taking place in the joint laboratory as quickly as possible.[4] Since all competitors equipped themselves with the same advanced technological capability, competition took place among semiconductor companies over which company would develop a new product most quickly. A cooperative environment enabled the producers to share technological information, stimulate innovation, and raise technological capability. But cooperation itself stimulated competition over the speed of product innovation using the shared advanced technology. In addition, each producer's introduction of products based on similar technological sophistication further stimulated market competition (Okada, 1999). Competition was taking place behind cooperation.

Similarly, most of the practices in interfirm cooperation involve diverse stringent measures to make those practices work effectively, though they undeniably require a strong sense of cooperation and coordination. For example, cooperative quality assurance programs require a high standard of quality. Any supplier failing to meet the standard is penalized by losing some part of its orders to other suppliers. Value analyses and engineering actually involve bonus payments to outside companies, and suppliers are

---

[4] Interview with Company Q manager.

monitored on how they seriously engage in these activities. Strategic pricing requires a strong sense of cooperation to reveal the detailed cost information to an outside company, and failure to do so and comply with a pricing plan results in a penalty or the termination of the business relationship. These cooperative practices are actually accompanied by monitoring, evaluation, penalties, or even the termination of the relationship. Cooperative practices are actually accompanied by some measures to introduce competition. Conversely, as the example of semiconductor sales suggests, competition can be the motivation behind cooperation; severe market competition compels sellers to develop cooperative relations, shifting the focus of competition from marketing to cooperative behavior.

Market efficiency, understood as generating a maximum total value of outputs from any given set of inputs (Shepherd, 1997), can be achieved through both allocative and non-allocative efficiency. Allocative efficiency is accomplished by allocating resources to maximize the output and non-allocative efficiency is determined by other factors such as developing group dynamics, soliciting commitment, and stimulating motivational drives. Allocative efficiency can be most effectively achieved by introducing competition and rewarding a winner. The market is the best known mechanism for harmonizing transacting actors to achieve allocative efficiency, but it is incapable of attaining non-allocative efficiency. In contrast, non-allocative efficiency is usually achieved by promoting cooperation and appealing to an individual's motivation. A single organization can be effective in bringing about a very high level of cooperation and stimulation to individuals, but it is quite weak in generating allocative efficiency.

Long-term relations are in many cases developed between two independent companies, hence influenced by the allocative-efficiency-achieving market mechanism. Yet, they involve cooperation and diverse cooperation-promoting measures that enable them to achieve non-allocative efficiency. They can involve practices to increase both types of efficiency by both creating a cooperative environment and introducing competition to cooperative relations or making use of market competition. This is the unique and dynamic mechanism of long-term relations. Neither market nor an organization can generate such dynamics by itself. In other words, the dynamics of Japanese companies' behavior cannot be understood in terms of either the intrafirm operations of suppliers, assemblers, and distributors or arm's length transactions. A significant part of their success should be attributed to the development of long-term interfirm relations.

These long-term relations involve at least some sense of cooperation and cooperative activities, since repeated interaction generates benefits like reduction in transaction costs and the mutual sharing of information and know-how. But developing long-term relations does not mean that interacting companies are not operating based on the rules of market competition. Responding to changes in corporate environments, a company can modify cooperative relations with a partner by introducing uncertainty and threat or even terminating the relationship. It can make use of market competition to influence the nature and characteristics of cooperative relations. Companies in developing long-term relationships have a range of options: to increase the degree of cooperation, to introduce severe competitive environments among cooperating partners, or to pursue a combination of these two strategies. In other words, long-term relations can delicately mix cooperation and competition. Then, how do such interfirm relations contribute to the dynamics of the Japanese semiconductor industry? This is the basic question pursued in this book.

Semiconductor companies, usually large in size, interact with diverse types of business partners. One category consists of intrafirm members, for example, semiconductor manufacturing units interacting with company-wide procurement and sales sections, research laboratories, or other manufacturing units within the company. Second, semiconductor producers interact in long-term relationships with two types of companies: some are large-sized with equal power and bargaining positions vis-á-vis semiconductor manufacturers, while some are small- and medium-sized and have less power and an inferior bargaining position. Interaction may also be with companies in the spot market, with little prospect for continued business transactions. Business transactions in the semiconductor industry take place in roughly four functional areas: (1) parts and material procurement, (2) manufacturing-equipment procurement, (3) sales, and (4) R&D. Each transaction may involve any of the four different types of partners. For example, a semiconductor manufacturing unit can purchase parts and materials from other manufacturing units of the same company, or from long-term suppliers (small-, medium-, or large-sized), or from companies operating in the spot market. This typology of partners also applies well to R&D. Many companies conduct internal R&D, and long-term partners engage in joint R&D. But technology and know-how can also be acquired by one-time transactions with an outside organization.

The diversity in business partner type and functional area undeniably generates complexity in a semiconductor company's operations. Yet, the selection of business partners is based on an interaction between their characteristics and the company's benefits; selection of a company's preference for one type of partner suggests the existence of more favorable advantages in interacting with them than with other partners. Then, what are the differences in the nature and characteristics of interaction by type of partner? What types of partners do semiconductor manufacturers choose for transactions in each functional area? To what extent do they choose the same types of partners for different functional areas? And what advantages does interacting with each type of partner hold? If long-term relations are the source of industrial dynamics, one advantage of these relationships lies in developing a set of expectations, rules, and norms of behavior for sustaining relations and generating mutual benefits. Then, what is the set of norms and values that is being developed by semiconductor manufacturers and their partners? Is it the same for all types of long-term partners?

Even minor partners may offer some benefits that other partners cannot contribute. And a semiconductor company is able to combine and coordinate benefits by interacting with different partners in different functional areas. How does an organization harmonize such complex intra- and interfirm relations? The answers to these questions, as addressed in this book, identify an overall structure of semiconductor manufacturers' interactions and help to illuminate their reasons for maintaining intra- and interfirm relations. The findings further our understanding of the sources of Japanese industrial dynamics and of the structure of the intra- and interfirm relations within the Japanese semiconductor industry.

If long-term relations are dynamic sources of the structure of the Japanese semiconductor industry, what is the mechanism that harmonizes cooperating partners? How does the harmonizing mechanism differ from that of intrafirm members or companies in the spot market? Long-term relations become an important device for enabling a company to delicately mix cooperation and competition, and they generate some advantages not found in spot or intrafirm transactions. This capability makes the relationships a key to explaining Japanese semiconductor companies' productivity and success in the golden age (Kinzley, 1991; Iwata, 1977; Pascale and Athos, 1981). This delicate mixture of cooperation and competition, a mixture of two concepts often considered contradictory, was developed out of a traditional strategy of rigid cooperation, which was in existence before the Oil Crises (Okada, 1993). Traditional and rigid cooperation in the past required a very strong commitment and tight reciprocity. Once special favors and privileges were given by one company to another, the recipient side was quite obliged to return the favor by responding to the demands of the company. As cooperative relations firmed up, both companies started feeling obliged to continuously provide favors and privileges. They came to strongly emphasize the importance of loyalty, obedience, obligation, and reciprocity.

In a sense, traditional and rigid cooperation placed primary importance on the normative aspects of relationships, making its content secondary. For any company, the predominance of relationships narrowly limited strategic alternatives, inhibiting its ability to adjust quickly to new market and techno-logical developments. The system was too rigid and inflexible to cope with the drastically- and unexpectedly-changing corporate environment. These limitations were strongly felt by many Japanese companies, especially in the late 1960s, when the rise of wages weakened their international competitive-ness in selling low-priced labor-intensive products. Efforts to correct this situation met resistance and indignation as a violation of established norms. For example, during the 1970s Oil Crises, large-sized Japanese companies terminated cooperative relations with many small- and medium-sized com-panies, and the latter openly expressed a sense of betrayal. These spurned companies were basing their criticism on a traditional and rigid sense of cooperation. The decade of the 1970s provided an excellent opportunity for Japanese companies to re-address the problem of normative restrictions (Okada, 1993).

To make the situation worse, the short life cycle of the semiconductor, severe competition over market share, and enormous R&D costs further pres-sured Japanese semiconductor companies to reconstruct cooperative inter-firm relations, without adopting market or intrafirm transactions. They trans-formed the traditional and rigid sense of cooperation into a more flexible one, which offered additional comparative advantage. They intensified co-operation by developing more cooperation-promoting measures, namely, pro-grams and practices that enhance joint or collaborative behavior for achiev-ing shared goals. To create flexibility in cooperative behavior, they also introduced competition-generating measures – programs and practices that stimulate competition, increase uncertainty, and create a threatening envi-ronment among business partners.

Competition-generating measures create competitive market environments for companies without long-term relations. For example, a company can openly distribute the specifications of required parts and materials to rel-evant suppliers, stimulate market competition among them, and select the lowest priced producer. If competition-generating measures are applied to long-term partners, a company first selects a limited number of cooperating partners, implements measures to stimulate competition among them, and consequently creates a competitive market-like environment. It is market-like, because behavior and the logic of competition become quite different from those in the competitive market once cooperative relations are involved. However, the number of companies involved becomes far less than the case of the market, since good cooperative relations can be developed with only a limited number of firms.

To create market-like environments, semiconductor manufacturers introduce several competition-generating measures: they bring the threat of instability to long-term partners in areas where cooperation is well established; they monitor and evaluate the performances of cooperating partners not only on objective outcomes, but more essentially on how beneficial and satisfactory cooperative relations are; and they stimulate competition among cooperating partners by rewarding with differentiated degrees of cooperative benefit and threatening possible termination of the relationships.

Figure IN.1 depicts the above processes in procurement relations. For example, Semiconductor Company A promotes cooperation with Suppliers P, Q, and R, while it also implements competition-generating measures with them. Semiconductor Company B does the same with Suppliers R, S, and T. In this case, Supplier R actually conducts business with both Company A and Company B. This is another competition-generating measure for Company A to force Supplier R to become financially independent as well as to enhance the latter's technological capability by interacting with another semiconductor manufacturer. But Supplier R's cooperative relations remain stronger with Company A than Company B, while it faces more stringent competition-generating measures from Company B than from Company A. Due to the implementation of competition-generating measures by Semiconductor

Figure IN.1 Cooperation-promoting and Competition-generating Measures in Procurement Relations

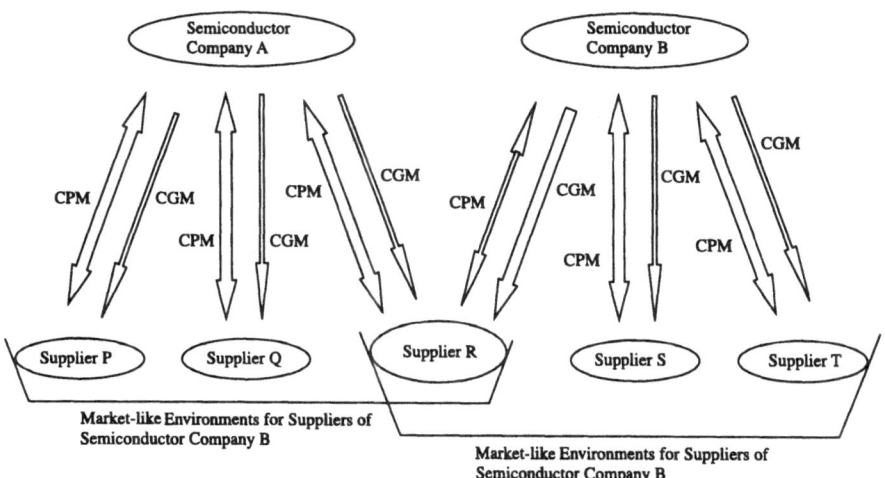

Note: CPM: Cooperation-promoting Measures; CGM: Competition-generating Measures

Company A, three suppliers to Company A come to compete among themselves. It is market-like rather than a true open market, because it is not an open competition where any company can come in, and because cooperative relations generate differences from regular market competition.

One well-known example is the multiple vendor system, under which semiconductor manufacturers use three or four suppliers for parts, with suppliers competing in terms of technological capability, quality, cost, delivery, and services. Any company, whether with long-term relations or not, has to compete in these aspects. But cooperating companies are monitored and examined on how strongly they are committed to present and future cooperative improvements of these aspects, how much they are willing to share resources and information in those improvement processes, and how much they try to meet expectations and generate synergetic benefits. For example, in the area of quality, the evaluation is not simply on product quality itself, but also on corporate processes such as the implementation of the quality assurance programs, their willingness to share information for solving problems, their risk-taking attitudes aimed at the quality of future products, and the seriousness and speed of a supplier's response to quality problems, suggestions, and improvements. Hence, competition-generating measures stimulate market-like competition among long-term partners in refined and intensified ways.

Cooperation-promoting and competition-generating measures are, in essence, contradictory to one another. As Table IN.1 shows, if a company tries to maximize the effectiveness of cooperation-promoting measures, the most suitable type of harmonizing mechanism of diverse actors (governance) is hierarchy. Actors inside an organization cooperate most effectively under the bureaucratic mechanism of control and coordination. And traditional-and-rigid-cooperation-based long-term relations follow, since tightly-knit and normatively-controlled interpersonal relations can also solicit cooperation effectively. Such tightness in control and coordination makes the mechanism unfit to create competitive environments, since competition can be more easily promoted in environments with low levels of control and coordination. In contrast, the lack of organizational control and normative restrictions among companies in the market allow companies to act more freely and engage in competitive behavior, which is unfit for generating cooperation. The market mechanism is the most effective for generating competition, while the least effective for implementing cooperation-promoting measures.

To effectively implement both cooperation-promoting and competition-generating measures, companies in long-term relations must sacrifice some cooperation, legitimize competition without destroying cooperative relations, and transform the nature of cooperation. They must nurture a sense of flexible cooperation that can accommodate these contradictory measures, whose effectiveness are reduced to medium and medium + levels in contrast with the other types of harmonizing mechanisms (Dore, 1986).

Flexible cooperation is primarily based on cooperative bargaining that heavily weighs the mutual interests and gains of interacting partners, though the influence of traditional thinking remains to some extent (Aoki, 1984). It strongly emphasizes higher joint gains by conducting business repeatedly, reducing transaction costs, enhancing mutual support, sharing necessary information, developing a sense of trust and cooperation, and soliciting commitment to relations based on the potential benefits of exploiting mutual interests. As Japanese companies implemented flexible cooperation, interfirm relations came to be bound more by mutual gains than by moral obligations, and the sense of obligation and reciprocity weakened, while commitment based on mutual gains grew stronger. Cooperation became more fluid, allowing partners to accept competition-generating measures, monitoring and

Table IN.1   Harmonizing Mechanism for Effectiveness of Cooperation-promoting and Competition-generating Measures

HARMONIZING MECHANISM		EFFECTIVENESS OF COOPERATION-PROMOTING MEASURES	EFFECTIVENESS OF COMPETITION-GENERATING MEASURES
Hierarchy (Intrafirm Relations)		High +	Low
Long-term Interfirm Relations	Traditional And Rigid Cooperation Based	High	Low +
	Flexible Cooperation Based	Medium / Medium +	Medium / Medium +
Market		Low	High +

evaluation and even to search for alternative interacting methods and partners, while enhanced cooperation became a means of ascertaining and increasing the potential for mutual gains. Transforming a sense of cooperation from traditional and rigid to flexible required the acceptance of contradictory elements in interactive behavior, while the importance of cooperative behavior remained as high as it was before.

I call this type of repeated interfirm interaction that results in a sense of flexible cooperation and allows the implementation of two basically contradictory measures competitive-cum-cooperative (CCC) interfirm relations. CCC relations allow companies to emphasize cooperation at some times, but at other times to emphasize competition among selected partners. A delicate mixture of cooperation and competition became a quite viable means to cope with uncertainty and generate interfirm flexibility and dynamics.

CCC interfirm relations, however, inherently face the danger of unbalancing the proper mix. To minimize this danger, the system requires a set of norms and values that sustains continuous relationships. I call it the logic of continuity. As conditions for continuity, CCC interfirm relations emphasize the importance of maintaining a sense of trust and cooperation, mutually sharing information and capability for survival, meeting rigorous expectations mutually and continuously, and developing future-oriented and risk-tasking activities. To make sure that such conditions apply, monitoring and evaluation are justified as a legitimate behavior, not as a threat to cooperation and trust. It is only after lengthy mutual efforts to solve problems and meet criteria enforced by competition-generating measures, that a partner could terminate the relationship. In a sense, only when cooperation-promoting measures become ineffective, and when either partner becomes non-responsive to competition-generation measures, is termination of the relationship acceptable as legitimate.

When a set of actors accepts the contradictory elements of cooperation-promoting and competition-generating measures as means for harmonizing actors, and behaves according to norms and values specified in the logic of continuity, CCC interfirm relations evolve into a CCC governance. It is a way diverse actors are harmonized and coordinated on a long-term basis by making use of a delicate mixture of cooperation and competition.

The logic of continuity is embedded in the Japanese institutional inheritance of interpersonalism, which, in essence, requires one to understand the expectations of others and to adjust oneself in accordance with their expectations rather than to act according to a self-centered evaluation of a situation. Long-term relations are the outcome of the path-dependent development of Japanese institutions modified to suit to turbulent corporate environments.

As transaction cost economics suggests, if fitness between governance and contingencies is a vital source of efficiency and dynamics, then the dynamics of the Japanese semiconductor industry is found in a match between CCC governance and the path-dependent institutional development of Japanese interpersonalism.

In Chapter One, I introduce the theoretical framework for analyzing intra- and interfirm relations and further elaborate the concept of CCC interfirm relations. To provide background for the empirical study, in Chapter Two, I examine the history of competitiveness in the Japanese semiconductor market, especially in relation to technological development and corporate strategy. Drastic changes in the nature of domestic and international competition, which were driven by technological innovation, have imposed difficult strategic choices for Japanese companies.

Chapters Three and Four introduce the empirical study. Chapter Three examines, first of all, whether long-term relations are actually dominant in all functional areas. Long-term relationships may be found predominantly in the functional areas of sales and the procurement of parts, materials, and equipment, since these transactions involve intensive interfirm interaction. In contrast, technological secrecy often prevents interfirm cooperation in R&D activities and makes intrafirm relations important. Secondly, focusing on the aspects of cooperation and power relationships, the chapter examines the

Table IN.2 Top Twelve Japanese Semiconductor Companies for Selected Years

COMPANY	1990 RANK	1990 SALES (BILLION YEN)	1989 RANK	1989 SALES (BILLION YEN)	1987 RANK	1982 RANK
NEC	1	725.0	1	680.0	1	1
Toshiba	2	700.0	2	660.0	2	3
Hitachi	3	580.0	3	540.0	3	2
Fujitsu	4	415.0	4	413.0	4	6
Mitsubishi Electric	5	375.0	5	341.0	6	5
Matsushita Electronics	6	280.0	6	255.0	5	4
Sanyo Electric	7	195.0	7	194.0	7	7
Sharp	8	190.0	8	180.0	8	8
SONY	9	160.0	9	140.0	10	15
Rohm	10	141.6	11	122.0	11	10
Oki Electric	11	136.0	10	136.0	9	9
Seiko Epson	12	90.0	12	70.0	14	

Sources: Sangyo Times, *Handotai sangyo keikaku soran* (Comprehensive list of semiconductor industry planning), (Tokyo: Sangyo Times, 1991), 336.
Sangyo Times, *Handotai sangyo keikaku soran* (Comprehensive list of semiconductor industry planning), (Tokyo: Sangyo Times, 1990), 2.
Yano Research Institute Ltd., The Japanese semiconductor and IC industry, (Tokyo: Yano Research Institute, 1984), 2.

nature and characteristics of intra- and interfirm relations. Chapter Four and Chapter Five examine the effects of cooperation-promoting and competition-generating measures and try to identify the reasons why semiconductor companies chose each type of partner in each functional area.

This study focuses on the top ten Japanese semiconductor companies in 1989 which were responsible for 85.8 percent of total semiconductor sales in Japan (Sangyo Times, 1990:2).[5] These companies are listed in Table IN.2. Since they dominated such a high percentage in the semiconductor market, other companies were considered less important. The small number of companies generates some problems in data presentation and statistical analyses, but these problems cannot be avoided due to the very limited number of large and successful semiconductor manufacturers. Seven companies out of ten, with 59.3 percent of the Japanese semiconductor market, agreed to answer questionnaires and interviews. One company agreed only to interviews, and two companies declined to cooperate. Twenty-nine interviews were conducted.

In my field research, I met with managers in the semiconductor manufacturing or public relations section, explained the purpose of the research, and left questionnaires for completion. Four types of questionnaires were distributed to each company. Questionnaires covered four functional areas: (1) the procurement of parts and materials; (2) the procurement of semiconductor manufacturing equipment; (3) sales; and (4) R&D. Because of the small number of semiconductor manufacturers, pretesting of questionnaires was conducted with two electronics-related high-technology companies that produce optical electronic goods, and photocopying machines and computers. Later, intensive interviews with a few members of each company were conducted about the questionnaire material and other issues.

---

[5] Unless specified, the semiconductor refers to any type, namely, the discrete or compound semiconductor, the monolithic integrated circuit (IC) that consists of two types of the MOS IC and bipolar IC, linear IC, or hybrid IC (for details, see Appendix Table 2.1). I focused more on the memory IC than other types. This is because (1) Japanese companies had the largest world market share (23.3% in 1990), (2) the most intensive development of more advanced process technology in Japan has been taking place in this field, and (3) Japanese success in this field led to the world leading position.

Chapter One

Theory of Competitive-*cum*-cooperative (CCC) Interfirm Governance

Many scholars attribute the dynamics of Japanese companies to the culture of cooperation and trust. Factors often mentioned include harmonious management, industrial relations, and government-business relations (Abegglen, 1984; Iwata, 1977, 1982, 1984; Cole, 1979; Tsuda, 1984; Johnson, 1982; Okimoto, 1989 Morishima, 1984; Murayama, 1982; Pascale and Athos, 1981). Some have heavily weighed the importance of harmonious interfirm relations such as long-term relations and Japanese horizontal and vertical *keiretsu*[1] (Gerlach, 1989; Inoue, 1985; Stowsky, 1989; Nakatani, 1984; Uekusa, 1987; Kinzley, 1991).

Long-term relations are clearly one of the most important factors that have contributed to Japanese companies' dynamics. Without them, some of the country's most highly valued practices would not exist. For example, Japanese companies are well known for engaging in just-in-time production that requires parts and material producers to deliver an exact number of products right before production or distribution. Quality assurance programs enable partner companies to improve quality and production processes, while strategic pricing allows them to plan for reducing production costs (Imai, M., 1986; Ishikawa, 1985). These practices greatly depend on a strong sense of cooperation and trust, cultivated through long-term relations.

Examining these practices, I have also found the existence of quite stringent environments, which make such cooperative practices work effectively. Any company that fails to deliver parts and materials just in time faces a

---

[1] Horizontal *keiretsu* means a structure based on power-symmetric interfirm relations, in which loosely-affiliated large-sized companies of diverse complementary backgrounds cooperate with each other in business activities. Vertical *keiretsu* means a structure based on power-asymmetric interfirm relations in which, on the one hand, large-sized companies, and on the other hand, small- and medium-sized companies closely cooperate with each other in business activities.

serious penalty, and another company may be introduced to cover the portion of the reduced order as a penalty. The same environment exists for companies that cannot meet very strict and high quality standards and cannot implement strategic pricing. Even the few examples mentioned here clearly suggest that long-term relations remain as an important source of Japanese companies' dynamics, because they delicately mix cooperation and measures that introduce competitive environments among cooperating partners. How did this dynamic form of long-term relations come to develop? Is the development related to cultural values and norms?

To answer the above questions two theoretical frameworks, Williamson's transaction cost economics and North's path-dependent institutional development, are used (Williamson, 1975, 1981, 1985; North, 1989, 1990, 1993). If the development of long-term relations is mainly due to the process of economic development or economic conditions, then Williamson's framework would solely explain why long-term relations exist. If the existence and dynamics of long-term relations in Japan are greatly owing to the cultural norms and their transformations of cooperation, then North's theoretical framework of path-dependent institutional development provides a better explanation. If so, the two approaches can be also integrated by treating North's path dependent institutions as a part of contingencies to influence a company's choice of long-term relations. To analyze Japanese long-term relations, it is useful to draw on and integrate these two frameworks.

In the rest of this chapter, I examine whether long-term relations can be explained by transaction cost economics. Then, identifying the inadequate explanation of the transaction cost economics, I develop a framework built on both Williamson and North. Using the concept of institutions, I shall briefly trace how the path-dependent characteristics of Japanese culture shaped the environment for maintaining interfirm cooperation. On these bases, I explain the mechanism of Japanese interfirm relations in terms of competition-*cum*-cooperation and develop propositions for empirical inquiry based on the combined framework.

A.   Theoretical Bases: Transaction Cost Economics and Path-dependent Institutional Development

Williamson's (1975, 1981, 1985) transaction cost economics opened an avenue of institutional inquiry beyond the narrowly confined realm of either the market (demand and supply under neo-classical economics) or the hierarchy (the theory of the firm). His focus on transactions, manifested in contracts, enables us to interpret market and hierarchy as two extreme forms of harmonizing diverse economic actors. Market, at one extreme, has no internalized transaction, while hierarchy, at the other extreme, has all transactions internalized. And intermediate forms, such as Japanese long-term relations, can be captured as forms that can mix characteristics of both market and hierarchy.

For some functions, a company may rely on internalized operations, while for other functions it may prefer to transact with companies in the market. The form of coordinating economic activities it uses is determined by decision-makers in order to minimize transaction costs based on the contingency of operations (Williamson, 1981). Using other forms increases the costs. Contingency factors can be opportunism, information impactedness (asymmetry of information between transacting partners), small number conditions (monopolistic or oligopolistic market conditions), uncertainty, bounded rationality, frequency of interaction, and asset specificity of product (the particular characteristics of product that link transacting partners). Williamson (1985) especially emphasizes the importance of frequency of interaction and the asset specificity of products in his later work. The choices of mechanism for harmonizing economic actors well fitted to contingencies are sources of efficiency and positive performances. This harmonization of transactions/ contingencies along with its coordinating mechanism has been termed governance structure by Williamson (1981).

With regard to Japanese cooperative intra- and interfirm relations, Williamson suggested the importance of cultural and social constraints on opportunism, and his argument propagated the use of terms such as "moral governance" (Williamson, 1985) and "trust governance" (Smitka, 1991). Morality and trust, as parts of norms and values, are important underlying traits of cooperative relations, and have significantly contributed to the formation of long-term relations. But they do not adequately explain why long-term relations become the dynamic sources of corporate operations.

Using the analytical tool of fitting firms' coordinating mechanisms and their contingencies, a group of theorists argued that subcontracting (a part of long-term relations in the case of Japan) is a phenomena that becomes conspicuous only at early or disintegrating stages of a firm or an industry. Attempts to reduce transaction costs in the market lead to internalized operations as firms grow, while attempts to reduce bureaucratic costs lead to the development of subcontracting at the disintegrating stage. As firms grow bigger, they merge and acquire smaller firms. Chandler (1977) argued that a firm operates with subcontracting relations at an early stage, but with its success it inevitably develops into an integrated large-sized company. Integrated mass production was the revolutionary means that enabled some companies to succeed in 19th century America. Following the arguments of Chandler and Shumpeter, Langlois and Robertson (1995) also asserted that recent needs for technological innovation compel companies to become large-sized, since transaction costs for acquiring technology from the market tend to be very high. Young (1928) argued that a firm, as it grows, vertically integrates subcontractors and internalizes diverse operations. But as these operations become too large, they spin off from the main company. Similarly, Markusen (1985) notes that as a firm grows bigger, organizational inefficiency leads to subcontracting relations.

None of their research is on Japan. But according to these arguments, Japanese long-term relations are a phenomenon of one stage of its economic development either an early stage or disintegrating stage. Attempts to reduce transaction costs should transform Japanese long-term relations toward more integrated mass production. Japanese companies, however, historically remained rather specialized (Fruin, 1992). The dual structure of large-sized companies and small- and medium-sized companies has persisted since the beginning of early Japanese industrialization. Furthermore, highly cooperative environments have made merger and acquisition in Japan extremely difficult to perform. This means that relationships between firms' choices of harmonizing mechanisms and economic contingencies found in the West do not explain well the continued existence of Japanese long-term relations. Explanation based on transaction cost economics is not helpful in understanding Japanese long-term relations. Its failure is because Williamson focused on contract-based transactions, which is the wrong focus in the context of Japan, and because a group of scholars explaining subcontracting have an inadequate understanding of Japanese cooperative environments.

These criticisms lead us to two questions. If we are trying to analyze the way intra- and interfirm actors are harmonized under different contingencies, what basic element, which still in the end results in cutting transaction costs, should we focus on instead of transactions? Which theoretical framework allows us to include cultural values and norms among the contingencies that influence the way intra- and interfirm actors are harmonized?

Williamson's focus on opportunism, transactions, and issues directly and indirectly related to contracts between companies ignores the importance of non-transaction-oriented activities for generating future transactions or even reducing opportunism (Okada, 1989a). For example, frequently visiting a customer, proving a seller's personal trustworthiness, and assessing his or her willingness to take future risks in cooperation with a customer often result in developing a long-term relationship in Japan. Accumulated tacit knowledge, shared values, and the coding and decoding capability of shared information cultivated between cooperating partners are also considered very important and effective in reducing transaction costs, even though they are not a part of contracts (Nonaka, 1989; Zajac and Olsen, 1993; Kogut and Zander, 1993). And sometimes when product-market conditions drastically change, the maintenance of a long-term relationship is valued more than short-term contracts or gain. In a sense, such behavior, sometimes without even involving contracts, may be perceived as long-term transaction-cost-minimizing for both sellers and buyers. Assuming that transaction-cost-minimization results in the dynamics of interfirm relations, then non-transaction-oriented activities are an important part of creating dynamic interfirm relations.

Arguing against the theory of transaction cost economics, supporters of the power and control approach shift their focus from transactions to bargaining power relationships and from contractual relations to coalition formations and power relations (Campbell and Lindberg, 1990; Pfeffer, 1987). This approach covers both contractual and non-contractual activities, and interprets intra- and interfirm relations as one of the variables that determine companies' bargaining power. Interfirm relations in a company's coalitions with and dependence on other companies determine the allocation of resources and rewards, and consequently influence performances (Hall, 1986).

However, due to its focus on power relationships, the approach emphasizes the degrees of control and dependency, the methods of control, conflicts, and power structure. It assumes a zero-sum game over the distribution of limited outcome. This assumption makes the approach incapable of illuminating Japanese cooperative behavior, since it involves a non-zero-sum game with shared goals and interests. The basic element that we need to focus on should be able to capture behavior based on both zero-sum and non-zero-sum games, since they differentiate capability to minimize transaction costs. This also means that the sharing and non-sharing of expected goals and interests between partners relate to minimizing transaction costs. But the focus on the non-zero-sum game does not deny the importance of power relationships, since symmetry or asymmetry between cooperating partners differentiates cooperative behavior.

Thus, the basic element used to interpret intra- and interfirm relations should accommodate a wide range of behavior, both transaction- and non-transaction-oriented, both normative and non-normative, both zero-sum-game- and non-zero-sum-game-based, and both goal-shared and goal-unshared. It should capture a variation caused by the differences in functional areas, power relationships, and product market conditions. The best comprehensive concept is interaction, the actions of two parties that influence each other's behavior. Semiconductor companies choose a type of interaction because of its effectiveness at minimizing transaction costs in the end. As will become clear, this focus on interaction is also a good fit with Japanese inherited culture.

Differences in the degrees of cooperation, interdependence, power relationships, frequency of interaction, asset specificity of products, and shared goals and interests greatly influence the nature and characteristics of intra- and interfirm interaction. The repeated use of a certain type of intra- or interfirm interaction institutionalizes a set of relations, measures, and practices, supported by a value system to maintain and justify the continuity. Eventually, actors come to harmonize their behavior under a certain type of intra- or interfirm governance.

Williamson provided a clear framework for understanding that fitting together contingencies and choices for harmonizing economic actors is a source of efficiency and positive performance. This analytical method is quite important for identifying the strength of long-term relations relative to the market or the hierarchy. But his cross-sectional analysis on economic conditions failed to include cultural norms and values as important parts of the field of contingencies. North's path-dependent institutional development solves this problem. He explained that institutions develop paths dependently and stimulate and restrict the behavior of actors within institutions to development in a certain direction. Such development functions to reduce transaction costs. If institutional inheritance is the source of fitness between contingencies and the ways actors are harmonized, it can be explained only by combining the two frameworks of North and Williamson. The next section explains the combined framework.

B.   Harmonizing Mechanisms and Path-dependent Institutional Development

Under what contingencies do the top executives of companies choose market or intrafirm transactions? And under what conditions do Japanese companies choose long-term relations? North has argued that such choices are path-dependent, strongly influenced by institutional backgrounds in which culture significantly influences the synthesis of diverse rules and norms that have been developed by other institutions. Hollingsworth and Boyer (1997) argued that each country, having different socio-economic, political, cultural, and geographical conditions, develops different institutional arrangements and consequently a social system of production that differentiates ways activities are conducted. They suggested that efficient forms of conducting business activities differ by country and even by sector (Hollingsworth, 1991). These arguments imply that existing national or sectoral conditions must have shaped the nature of the long-term relations which became a dynamic source of Japanese companies' performance.

This focus on path-dependent institutional development and on social systems of production can be enriched by Williamson's emphasis on governance, the harmonizing mechanism of diverse actors. These approaches, interpreting the fundamental function of governance and institutions as transaction cost minimization, need not be contradictory. Rather a combined framework can become an important device to analyze long-term relations, enriching understanding by having both a historical perspective and a cross-sectional analysis.

In the context of a company, both governance and institutional inheritance can function to minimize transaction costs. Minimization is achieved if there is the right form of governance for a certain function in a company to harmonize diverse related actors. The choice has to generate the best fit between governance and the existing contingencies of a company's functional area, with contingencies strongly influenced by diverse path-dependently-developed institutions, including culture (see Figure 1.1). Hence, to minimize transaction costs for a certain functional area a company has to develop a form of governance that fits well to contingencies strongly influenced by institutions. This new analytical framework, built on transaction

Figure 1.1   Harmonizing Mechanisms and Path-dependent Development of Institutions

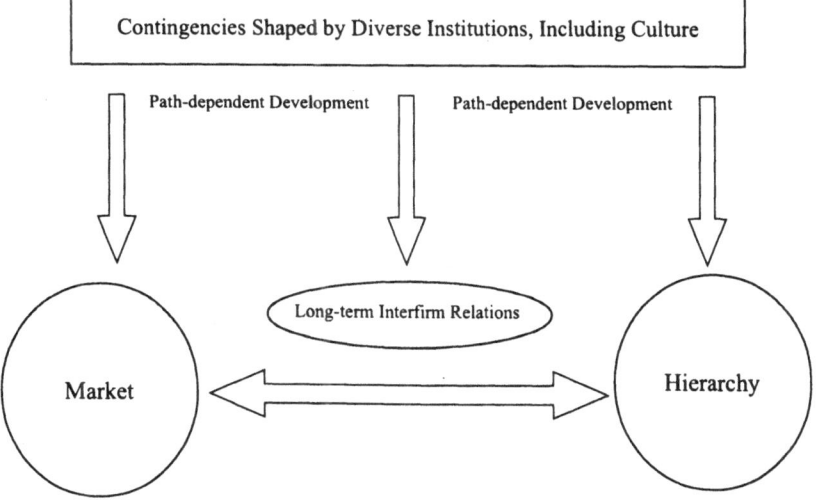

cost economics and path-dependent institutional development, can be used to identify the dynamics of CCC long-term relations.

1.   Roles of Institutions

Institutions restrict and stimulate individual and organizational behavior, and they also largely determine the development of systems and institutional arrangements. Institutions are "rules, enforcement characteristics of rules, and norms of behavior that structure repeated human interaction" (North, 1989:1321). They are the rules of the games of human interaction, consist of informal rules (social norms, codes of conduct, customs, etc.) and formal ones (constitutions, laws, etc.). In contrast, individuals and organizations are players in games (North, 1993).

Political institutions are not only the formal and informal rules established by the state (institution building), but also the enforcement characteristics (institution enforcement). Similarly, educational institutions are not only the formal and informal rules propagated through education, but also the way educational organizations enforce the rules. Culture, as a super-ordinate institution, is an integrated body of diverse norms and values developed by other institutions. It establishes deeply ingrained patterns of human behavior, and strongly influences other institutions such as political and educational ones. Based on the rules of the games developed and enforced by institutions, individuals develop forms of interpersonal relations, while organizations develop patterns of intra- and interorganizational relations. Under the influence of institutions, a set of organizational relations forms a system in each specific area, and a set of interlinkages among systems characterizes an institutional arrangement (Sjostrand, 1993:9).[2] Business activities, involving diverse institutions, are harmonized by a business system (Whitley, 1992).

In short, the rules of the games played out in repeated human interaction are created and enforced by diverse factors: (1) the existing overall traits of a given culture which is itself a super-ordinate institution that integrates the norms and values of diverse other institutions; (2) the development of other institutions such as political, educational, and financial ones, and (3) the development of interpersonal and organizational relations, a system in each specific area, and a set of interrelated systems under the influence of institutions. The development of long-term relations in Japan can be understood as

---

[2] An institutional arrangement is defined as a set of interrelated systems that regulates the behavior of individuals and organizations in recurrent situations in society.

the manifestation of institutions, systems, institutional arrangements, and the super-ordinate institution of culture.

Interfirm relations may simply be contractual ones, or they may involve moral commitments, identity gained from shared information, and information creation by participation and interactive learning (Aoki, 1988; Cremer, 1990, Lundvall, 1988). The latter type of relations, namely, long-term relations, allows the development of a pre-committed set of rules designed to constrain and stimulate future interaction (Khalil, 1995). Even while their rules and norms can restrict actors' alternative choices and may also function to restrict the direction of development, long-term relations have a potential for developing new rules of human interaction, new systems in specific areas, new institutions, and consequently new institutional arrangements.

Institutional development becomes conspicuous, especially when changes in environments (historical incidents, market conditions, etc.) are drastic and require radical alterations in human interaction, systems, and institutional arrangements. But development is made in such a path-dependent way that it supports the continuity of past institutions, with a little alteration to accommodate the environmental changes (North, 1990). Such institutional developments in different countries generate differences in the transaction and production costs of economic activities, and consequently they affect economic performances (North, 1990).

Hence, North's contingencies function to restrict as well as to stimulate the development of intra- and interfirm relations in certain directions. And thus, within this framework, Japanese long-term relations are stimulated and restricted by institutional backgrounds. To investigate the dynamics of long-term relationships in the Japanese marketplace, researchers must develop typologies of the harmonizing mechanisms (governance) and interactions that take the range of influence that institutional background has into account.

2.   Types of Intra- and Interfirm Interaction and Governance

As specified in Figure 1.1, the two extreme forms of governance are market and hierarchy. Market governance involves spot interaction, in which each interaction, based on a zero-sum game, does not have any influence over the next one (see Table 1.1). Frequent occurrence of spot interaction results in developing a set of interfirm relations, practices, and measures suitable for market governance. Some concrete examples are classic contract-oriented transactions, such as self-liquidating sales, auctions, and discontinuous short-term contracts (Lindberg, et al., 1991).

The other extreme is hierarchical governance, under which a set of actors are harmonized by intrafirm relations, measures, and practices supported by the bureaucratic principles of control and coordination and an organization-specific value system. In many cases, internalized interaction in a company takes place between two different units within the same company. It could be between two divisions that operate as semi-independent profit centers,

Table 1.1 Types of Governance and Interaction

TYPES OF GOVERNANCE AND INTERACTION	EXAMPLES
A. Market Governance	
Spot Interaction	-self-liquidating sales
	-spot-market contracts
	-discontinuous short-term contracts
B. Competitive-*cum*-cooperative (CCC) Governance	
Horizontal CCC Interaction	-follow-on weapons consortiums
	-R&D alliances
	-franchise contracts
	-limited corporate interlocks to stabilize resource and capital flows
	-stable contracting
	-equity joint ventures
	-interfirm agreements
	-licensing
	-production alliances
	-Japanese public ventures' contracts with the private sector
	-long-term relationships between firms with symmetric capability
Vertical CCC Interaction	-subcontracting
	-long-term relationships between firms with asymmetric capability
C. Hierarchical Governance	
Internalized Interaction	-vertically-integrated companies
	-horizontally-integrated companies
	-conglomerates

Sources: Leon N. Lindberg, John L. Campbell, and J. Rogers Hollingsworth, "Economic governance and the analysis of structural change in the American economy," in The governance of the American economy, edited by John L. Campbell, J.Rogers Hollingsworth, and Leon N. Lindberg, (New York and Cambridge: Cambridge University Press, 1991).
Yoshitaka Okada, "*Nichibei handotai sangyo ni okeru shakai keizai tosei kozo no hikaku* (Comparison of socio-economic coordination structures in Japanese and U.S. semiconductor industries)," in *Kawariyuku Nihon no sangyo kozo* (Changing Japanese industrial structure), edited by Akinori Marumo, (Tokyo: The Japan Times, 1989), 52-98.

between a semiconductor division and research laboratories, or even between a semiconductor division and a subsidiary of the same company. Some concrete examples are vertically and horizontally integrated companies, and conglomerates.

Long-term relations are between these extremes and there is complexity especially in the Japanese case. In the 1970s, facing the Oil Crises, Japanese companies introduced diverse competition-generating measures to make traditional practices more flexible and open. They introduced competition, uncertainty, and threats to cooperative interfirm relations. Such deliberate introduction of competition-generating measures is often found in procurement relationships. In contrast, severe market competition itself provides a competitive environment in the functional area of sales. Even so, cooperation is still pursued as an important means of winning competition.

Long-term relations involve competitive-*cum*-cooperative (CCC) interaction, a set of mutually influencing actions by long-term business partners that delicately mixes both cooperation and competition. When a set of actors accepts the contradictory elements of cooperation and competition and follows norms and values developed in cumulated interaction, their behavior becomes harmonized under CCC governance.

CCC interaction can be horizontal and vertical. Large-sized companies, having similar levels of technological and managerial capabilities, comparable bargaining power, and equal size, engage in horizontal CCC interaction with semiconductor manufacturers, while small- and medium-sized companies, having inferior capabilities, weaker bargaining power, and smaller size, engage in vertical CCC interaction.

Examples of horizontal CCC interaction are limited corporate interlocks to stabilize resource and capital flows, stable contracting and continuous business transactions (Lindberg, et al., 1991). Alliances between companies of complementary technology or between competitors are important. The exchange of information and know-how in complementary areas is vital for expanding sales channels, generating a scale economy, and speeding up innovations for winning competition. Examples of vertical CCC interaction include subcontracting systems, vertical *keiretsu,* or long-term relationships between large-sized companies and small- or medium-sized companies (Yoshihara, 1979; Patrick and Rosovsky, 1976; Allen, 1981; Ueda, 1978). Each interaction (or governance) develops its own practices and measures, given different contingencies (functional areas, market conditions, power relationships, and other characteristics of partners). These typologies enable us to analyze the conditions under which each type of interaction takes place.

C.   Institutions and Transformation of Cooperative Behavior

1.   Traditional and Rigid Cooperation

Human-relations-oriented social norms are considered highly influential in structuring human interaction in Japan. These norms greatly influence the rule making and enforcing actions of the state, the content of basic education, and the ways inter- and intrafirm relations are established. Such institutional factors restrict and stimulate the transformation of individuals' concepts of cooperation, especially when corporate environments[3] are turbulent (Fruin, 1992; Hollingsworth, 1991; Whitley, 1992; North, 1990, 1993; Williamson, 1975, 1985). Since the transformation of cooperative behavior from traditional and rigid to flexible enabled Japanese companies to develop competitive-*cum*-cooperative interfirm relations, I shall briefly trace the history, showing how such behavioral change became possible and what environmental changes took place.

The basic orientation of Japanese society and culture can be described as interpersonalism (*kanjinshugi*) rather than individualism (*kojinshugi*).[4] Interpersonalism is the psychological process of directing one's behavior, to give more importance to relationships with others than to the internal self.[5] It is important in interpersonalism to understand the expectations of others and to adjust oneself in accordance with their expectations rather than to act on self-centered evaluations of situations. Cooperation in Japan is the manifested behavior of interpersonalism, directed toward goals based on common interests and mutual expectations. Interpersonalism is the underlying logic and environment that supports cooperative behavior. Trust between partners is required and betrayal of expectations, if correction efforts fail, often results in the termination of interpersonal relationships. Such norms of

---

[3] Corporate environments in this book mainly refer to either market conditions or incidents that significantly influence market conditions.

[4] See Hamaguchi (1977) and the Masuda Foundation (1992). Members in a Masuda Foundation research project translated *kanjinshugi* as contextualism rather than interpersonalism. I have chosen the latter translation, since it emphasizes more strongly the importance of relationships, while the Chinese characters literally mean the latter.

[5] See Hamaguchi (1977) and Befu (1986:22). Benedict (1969) points out that Japanese behavior is restricted by a very complex system of external pressures and interpersonal relations, and that the Japanese lack the most important principle in human life, namely, self-initiating behavior. Similarly, Nakamura (1968) argues that Japanese Shintoism dramatically influenced the ways that Buddhism and Confucianism were interpreted in Japan, and weakened the importance of universally applicable moral principles emphasized in Buddhism and Confucianism.

interpersonalism are strongly institutionalized in Japan,[6] influencing even economic interactions and cultivating a ground for Japanese business practices (Imai and Kaneko, 1988).

According to Bellah (1957), the origin of interpersonalism lies in the concept of mutual reciprocity – a practice of or a belief in blessing (*on*) and return to the blessing (*hoon*) – that comes from a combination of Confucianism, Buddhism, and Shintoism. In the Tokugawa period (1603-1868), reciprocity was seen as unlimited, since the initial receiver of the deity's blessing, the ruler, could entitle himself to receive an unlimited amount of reciprocity from subjects in return for his blessing. But in interpersonal relations where the ideology of unlimited reciprocity was not applicable, it was important to give an equal amount in return for kindness or favors that one received (*giri*). For example, profit-making activities came to be increasingly justified in the Tokugawa period as morally acceptable behavior, but only when merchants followed the rule of limited reciprocity called *jiri-jita* (one can make profits only when one gives equal benefits to others) (Takahashi, 1969). Profit making was justified as cooperative behavior by satisfying the mutual expectations of partners. This is one reason why Japanese companies, from an early period of their development, have emphasized the importance of serving society and customers.

The norms and values of the Tokugawa period were transformed in the Meiji Period (1868-1912) by major historical events and changes in educational and other institutions. The newly introduced Western-style educational institutions stressed the acquisition of Western intellectual knowledge and the growth of individuals, and this was a change from the previous emphasis on the virtues of loyalty and self-sacrifice. The western influence caused dissatisfaction among people and teachers (half of whom were former *samurai*), and generated revolts against the government. Responding to such movements, the government started teaching morality courses based on Japanese traditional ethics in basic education. These courses emphasized the importance of education for cultivating social morality and the significance of Western intellectual knowledge as a means of contributing oneself to national development, rather than as a means of individual growth or self-enrichment. Indicating that education was a direct order from the Emperor, the government appealed to loyalty to the Emperor and successfully propagated

---

[6] Emphasis on interpersonal relationships seems to be rather unique to Japan, since mutual dependence and cooperation in China mainly remain around family relationships and those in Korea tend to be quite weak (Whitley, 1992:174, 186, 197).

the idea that educational achievement was important (Passin, 1967; Nagai, 1971). Hence, education preserved some traditional sense of unlimited reciprocity to sustain the absolute power of the Emperor (*chu*), the authoritarian power of the head of the household (*ko*), and the respect due to teachers (*shi*) (Benedict, 1969; Tokyo Daigaku Shakaikagaku Kenkyusho, 1992; Nakane, 1970; Minami, 1955, 1980). Under the traditional sense of reciprocity, cooperative behavior of people vis-á-vis feudalistic power holders meant self-sacrifice, self-denial, complete subjugation, blind respect, loyalty, and obedience. It was extremely tight and rigid, and there was hardly any room for individual initiatives. New Western knowledge and newly established Western-style institutions, however, gradually influenced people to think in terms of limited reciprocity. Especially in the business sector, the introduction of the factory system and the expansion of market activities in the Meiji period made it less easy to maintain feudalistic relations between an employer and employees.[7] Besides, labor disputes in the early industrialization period forced owners and business elite to defuse these problems, not by appealing to the feudalistic subjugation of workers, but by strongly showing the paternalistic benevolence of employers and emphasizing harmony between labor and management (Kinzley, 1991).

For example, in order to defuse militant activities among workers, companies came to offer some favorable conditions for workers such as the bonus system, nurseries and dormitories, thereby introducing the welfare-oriented aspects of the Japanese *ie* institution in industrial activities (Hazama, 1984). These conditions were interpreted as the employer's offering of paternalistic favors and solicitation of favorable returns from employees. But since employers quite clearly retained their power and authority, and since the layoffs of workers were frequent, workers constantly faced uncertainty with respect to their positions and felt a strong sense of hierarchical inequality.

Feudalistic absolute subjugation was eventually replaced by hierarchical control with a sense of limited reciprocity on both employer and employee sides. I shall call such norms and values hierarchy-oriented interpersonalism.

---

[7] Before the introduction of the factory system, the contract work system was predominant. A firm contracted out production to masters (*oyakata*), some of whom used contractor's production facilities, while some used their own workplaces. But relationships between an employer and direct employees (*shokuin* and some *oyakata*) and between masters (*oyakata*) and their subordinates (*shokunin* and *totei*) in manufacturing activities were bonded by feudalistic interpersonalism with a strong sense of unlimited reciprocity (Hazama, 1984:33-35). There was hardly any freedom on the part of *totei* to even express their complaints.

Workers' cooperation with owners and managers meant returns to paternalistic benevolence and subjugation to power holders, while they still felt the need for some self-sacrifice, obedience, and loyalty, the remnants of interpersonalism in the Tokugawa period. Individual initiatives, freedom, and expressions of personal interests, including contradictory attitudes, were accepted to some extent, but only within rigidly controlled power relationships. Institutional transformation was made in a way that increased the level of worker participation in the name of harmony and cooperation. And the logic of mutual expectations under hierarchical interpersonalism facilitated the participation of workers as well as the preservation of power positions of owners and managers.

The end of World War II brought another period of drastic transformation. To democratize and demilitarize Japan, the Supreme Commander of Allied Powers (SCAP) dismantled the pre-war hierarchical, paternalistic, and undemocratic institutions such as *zaibatsu*, the Imperial Household, *ie*, and the existing structure of educational institutions. The new principles of democracy and equality contradicted the old norm of hierarchy-oriented interpersonalism and created a sense of anomie among the Japanese (Moos, 1975). In the face of this denial of past social norms, the Japanese started searching for new interpretations of interpersonalism, mixing some traditional norms with newly propagated ones, such as equality and democracy. And they developed norms that could be most suitably applied to a new contingency.

Overcoming a period of sharp criticism of traditional norms in the 1950s and 1960s,[8] a group of scholars started advancing a new interpretation. In her study of Japanese society, Nakane (1970) showed that the Japanese had begun to display a group orientation based on *ba* (particular relationships which bind a set of individuals into one group) with an emphasis more on equality among group members than on hierarchical and authoritarian relations. They also began to stress the importance of situational positions in particular frames rather than universal attributes. To be a member of a group required total involvement and commitment to a group, not just functional participation. And such involvement and commitment differentiated the intensity of interactions and reciprocal relationships, the flow of information, and benefits from cooperative activities between insiders and outsiders. A company came to be the most important group for men to show total in-

---

[8] See Maruyama (1961), Otsuka (1973), Kawashima (1950), Minami (1955), and Nakamura (1968).

volvement and commitment. One significant difference, however, was that tight group control stimulated competition vis-á-vis other groups, creating an environment for group initiatives without much incentive for individual initiatives.

Such behavior was enriched and enhanced through basic education. Most grade school students were organized in classroom units, each with a strong practice of internal cooperation and a sense of competition vis-á-vis members of other classroom units (White, 1988). Hence, the post-World War II system of basic education functioned to spread a sense of cooperation within a group, but competition vis-á-vis other groups. Some mixing of cooperation and competition started taking place at this period.

Again, a particular human nexus and intensive reciprocal relationships were considered important, suggesting the continuity of interpersonalism as the primary core of social norms. I shall call this institution group-oriented interpersonalism. Cooperation took place under the environment of total commitment and involvement with a group, harmony with other group members, and group decisions and responsibilities. Individuals' behaviors were tightly harmonized within a group environment. The SCAP's denial of old feudalistic and hierarchical values and its introduction of the concept of equality and democracy drastically changed the content of cooperative behavior. But even so, those new norms and values were propagated on the old logic of interpersonalism, resulting in the development of a strong group orientation.

A group of management scholars, focusing narrowly on the positive contributions of Japanese values and tradition, argued that the development of long-term interfirm relations is greatly owing to the culture of cooperation and trust (Morishima, 1984; Iwata, 1977; Tsuda, 1984; Lebra, 1976; Murayama, 1982; Bellah, 1957; Pascale and Athos, 1981). These researchers, however, failed to understand that over-cooperation tends to mutually restrict individuals' behavior, makes interfirm relations rigid, and reduces flexibility and dynamics (Granovetter, 1985). Another institutional transformation became necessary to help industries cope with turbulent corporate environments.

2.   Transformation to Flexible Cooperation

Group-oriented interpersonalism faced difficulties in the 1970s and 1980s. Higher labor costs, the Oil Crises, trade disputes, the multinationalization of Japanese companies, and the needs for technological innovations and speedy market activities made group-oriented interpersonalism difficult. Traditional and rigid cooperation placed primary importance on the normative aspects of relations, leaving the content secondary, and thus narrowly limited strate-

gic alternatives for a company to adjust quickly to new market and technological developments.

These limitations were strongly felt by many Japanese companies when
the rise of wages weakened their ability to sell low-priced labor-intensive
products in the late 1960s, thus lowering their international competitiveness.
Attempts to correct this situation faced resistance and indignation as viola-
tions of group-oriented norms. During the Oil Crises in the 1970s, large-
sized Japanese companies terminated cooperative relations with many small-
and medium-sized companies, the latter openly expressed a sense of betrayal
that the former did not keep their obligations, loyalty and reciprocity. Terminated companies were basing their criticisms on group-oriented
interpersonalism.

The 1970s provided an excellent opportunity for Japanese companies to
readdress the problem of normative restrictions and institutional transfor
mation (Okada, 1993). Since social norms are much slower and harder to
change, for the sake of survival, companies aggressively searched for new
forms of interfirm relations. They coped with such severe conditions without abandoning and replacing cooperative interfirm relations with market or
intrafirm transactions. They transformed their long-term relations into something much more flexible, open, and dynamic.

Interpreting these changes, a group of scholars argued that the Japanese
cleverly mixed traditional and modern behavior. Each studying a different
traditional practice, Dore (1987), Ouchi (1984), Okimoto (1989) and Calder
(1988) showed that the factors of trust and obligation, informal and diffuse
behavior, informal networks, and special reciprocal relationships mixed well
with modern business behavior. These authors recognized that traditional
norms and values played very important roles in helping the Japanese system to generate dynamics, and that the meaning of cooperation has been
transformed. But they explained the system by simply reinterpreting Japanese traditional values, and did not give proper importance to new innovative attempts to escape from some aspects of normative restrictions. They
did not explain how cooperation transformed and how companies made long-
term interfirm relations flexible and dynamic.

Recently, the dynamic aspects of cooperative relations have attracted more
attention. Granovetter (1985) and Powell (1990) have argued that socially
embedded networks are an important source of efficient business behavior.
Some scholars have valued cooperative interfirm relations as an important
source of generating efficiency in business activities. They have argued that
interfirm relations develop networks, cultivate trust well, generate a smoother
and more enriched flow of information, allow effective coordination, and

increase efficiency in operations (Aoki, 1984, 1988; Imai, K., 1990; Imai, K. and Kaneko, 1988, Sako, 1992). These networks become information nodes as well as mediating channels for diverse interests. This openness in knowledge and information provides important opportunities for mixing different elements of diverse cultures and societies and for creating new and rational values and institutional arrangements (Imai, K., 1990).

Despite such openness, Asada (1984) and Murakami (1987) warn that it is not Western individualism that is becoming prevalent. Young generations have lost loyalty to a group, but still show a strong conformity to interpersonal harmony rather than individualistic behavior (Murakami, 1987). New attempts to reform educational institutions started taking place in 1979. In 1983 the Conference on Culture and Education, considered the most powerful advisory body in education policies since the postwar educational reform, was established. A significant number of prominent businessmen, who saw the limitations of the old institutions in a new age, were designated as members. The group came to emphasize creativity, freedom of students, and social responsibility. Individualism was considered a poor fit for the Japanese context (Fukayama, et al., 1986). Just a few incidents suggest that a new form of interpersonalism was evolving, as individuals broke away from a group orientation to an open network that could accommodate anybody sharing similar interests. If cooperation is the manifested behavior of interpersonalism, then this transformation should also influence the meaning of cooperation.

Game theory better explains the nature and characteristics of cooperation. Aoki (1984: 61) defines a cooperative game as one "in which players can conclude a binding agreement as to what outcome will be chosen to exploit the possibility of common interests." Partners bargain and engage in non-zero-sum games, believing that a better performance in either company will result in higher joint gain (Chamberlain and Kuhn, 1965; Peterson and Tracy, 1988). Cooperative games require repeated interaction to benefit from enhanced mutual support, positive feedback, accurate communication and information exchange, commitment to work and productivity, and a sense of trust and cooperation. But in the Japanese context, a binding agreement strongly relies on a value system rather than Western-type contracts.

Unlike a traditional sense of rigid cooperation, cooperative bargaining allows a significant degree of flexibility in interfirm relations (Tjosvold, 1984). As a way of assuring mutual gains in a turbulent corporate environment, cooperative games even accept contradictory behavior to cooperation as a valuable means for enhancing mutual support and gains. Some examples are the monitoring and evaluation of benefits that interacting companies receive.

Fruin (1992) and Sabel (1994) emphasize that economic learning and monitoring are much more important than trust and cooperation. Economic learning means "acquiring knowledge to make and do the things valued in markets" (Sabel, 1994). Sabel even argues that trust and cooperation are the outcome of economic learning and monitoring. In the case of Japan, in general, a normative value of interpersonalism and cooperation weighs much heavier than learning and monitoring. But even so, learning and monitoring result in generating competition among cooperating partners, even to the point of threatening ill-performing companies with terminating cooperative relations, and thus making traditional and rigid cooperation more flexible.

Besides, introducing openness and fluidity in cooperative activities results in accepting previously non-cooperating companies as partners. This type of collaboration, though it differs from interaction with a long-term cooperative partner, allows cooperative interfirm relations far more flexible than traditional cases.

In other words, cooperative behavior came to be more open, fluid, and flexible, bound by more mutual gains than moral obligations. The sense of obligation and reciprocity was weakened, while commitment based on mutual gains was strengthened. Yet, even though the focus on mutual gains became the major source of cooperative dynamics, it is but important here to recognize the path-dependent nature of cooperation. Cooperative behavior is still sustained by the cultural values and norms of interpersonalism, not simply rational calculation. This is one reason why cooperative long-term relations persist, and function and achieve differently from corporate relationships in many other parts of the world. Hence, cooperative behavior basically remains as important as it has been for many decades. And this is why long-term relationships have not disappeared and will not disappear.

Traditional and rigid cooperation are different from flexible cooperation, a system that cultivates a flexible environment for actors to accept even contradictory elements. The development of network-oriented interpersonalism as a contingency and flexible cooperation as its manifested behavior were the base of CCC interfirm relations.

D.   Competitive-*cum*-cooperative Interfirm Relations

Starting from the late 1960s, Japanese companies in general faced radical changes in corporate environments, such as higher labor costs, the Oil Crises, catch-up with Western technology, trade disputes, and the multinationalization of Japanese companies. For Japanese semiconductor manufacturers, the situation was far worse. Being behind the U.S. in memory integrated circuit technology from the early 1960s, superior and lower-priced U.S. integrated circuits came to dominate the Japanese market. Moreover, a giant semiconductor manufacturer, Texas Instruments, had established a joint venture with SONY in Japan in 1968. For Japanese semiconductor manufacturers, the market for advanced products, where their future lay, was in a very problematic condition.

This situation resulted from the nature of technological advancement in the semiconductor industry in general. It occurred because: (1) the life cycle of a semiconductor is short; (2) semiconductor technology quickly changes; (3) companies compete for a bigger share of a product market to cover the high cost of initial investment; and (4) an enormous sum of R&D fund is necessary to keep winning the technological race (Nikkei Business, 1984a, 1984b). Japanese semiconductor manufacturers were especially hard pressed, since conditions in the semiconductor industry had been extremely turbulent. There was urgent need for the transformation of interfirm cooperation to something much more flexible, fluid, and open. As a cumulative outcome to cope with such difficult market conditions, the competitive-*cum*-cooperative interfirm relations evolved.

1.   Cooperation-promoting and Competition-generating Measures

In repeated interaction, companies develop an understanding of the way transactions are conducted, share information, accumulate know-how about each other, cut transaction costs, nurture at the least some sense of cooperation and trust, and construct a cooperative environment. But long-term relationships, in many cases, develop between independent companies operating in the market without involving organizational integration. Interacting companies basically remain free to modify their interfirm relations in response to changes in corporate environment. Hence, long-term relations accommodate diverse options. On the one hand, a limited range of cooperation at the initial stage can be further extended by promoting good coordination in a wide range of operations, sharing a norm of satisfying mutual expectations for higher achievement, and even jointly developing future-oriented activities. On the other hand, long-term relations can be made stricter by introducing uncertainty and a threat to cooperative relationship. In the worst case, interfirm relations can even be terminated. Such measures function effectively to create openness, flexibility, and fluidity in a tightly knit relationship. Many companies opt to further enhance cooperative relations, while introducing competition, uncertainty and threat whereby they instigate a market-like environment. Mixing cooperation and competition can create some advantages hardly generated in spot or intrafirm transactions. This mix is the key to explain the productivity and success of the Japanese semiconductor industry in its golden age (1977-1992).

In order to amalgamate cooperation and competition, two types of measures are very important. One is cooperation-promoting measures, that enhance joint or collaborative behavior and help achieve shared goals. Some examples of such programs and practices are: mutual assistance in finance, personnel, procurement, and training; mutual information exchange on market, technology, and customers; resource interdependence for joint procurement and coordinated production; and innovation-stimulating activities conducted through diverse practices including joint R&D. (Details will be given later in this chapter).

On the other hand, competition-generating measures are important for making cooperative relations more flexible, fluid, and open. They introduce competition among cooperating partners, increase uncertainty in the continuity of long-term relations, and create a market-like and threatening environment. Some examples of such programs and practices are: strategic pricing, introduction of the new supplier to a selected group of cooperating suppliers, flexible contracting, tough and continuous negotiations over price, contract termination, explicit penalty for poor performance, evaluation and factory-site inspection, etc. Some of these measures are applicable to both market transactions and long-term relationships. (Details will be provided later in this chapter).

Cooperation-promoting measures work effectively within a cooperative environment, while competition-generating measures do so in a competitive environment. Hence, the effective implementation of these two types of measures, in essence contradictory to one another, requires some modification in the nature of cooperation. Flexible cooperation has to be developed so that interacting partners accept some competition-generating measures without destroying the cooperative relationships, accommodate these contradictory measures as a legitimate part of interfirm behavior, and nurture them effectively enough to generate a different dynamic outcome (Dore, 1986). Hence, competitive-*cum*-cooperative (CCC) interfirm relations allow companies interchangeably to promote cooperation at some times and at other times to emphasize competition. Such a delicate blend of cooperation and competition, found only in long-term relationships, becomes a quite viable means of coping with uncertainty and fostering interfirm flexibility and dynamics (Asanuma, 1985; Yoshino and Lifson, 1986; Nakamura, 1989).

2.   Flexible-synergy and Market-like Effects

Cooperation-promoting measures enhance synergy effects, while competition-generating measures intensify market effects. Synergy effects are generated by cooperative behavior, inclusive of all types of cooperation, the rigid as well as flexible, and also bureaucratic coordination. They take place whether the cooperative behavior is the outcome of strong norms and values, or cooperative bargains, or results from bureaucratic control. Similarly, the introduction of competition, uncertainty, and threat generates market effects among interacting partners. The market mechanism is considered most effective in maximizing the severity of competition and uncertainty with the implicit understanding that the relationship may be terminated after each transaction.

CCC interaction, however, may create different types of effects ranging from simple cooperation to competition. Cooperation-promoting and competition-generating measures in CCC interaction result in flexible-synergy and market-like effects respectively. Flexible synergy effects are the outcome of flexible cooperation, where partners engage in cooperative bargaining to maximize mutual gain. Market-like effects are the outcome of introducing competition, uncertainty, and threat among interacting partners, in a context of flexible cooperation or organizational coordination. It means that CCC and internalized interaction by involving a cooperative environment can generate market-like effects better than a spot interaction that involves no cooperative environment.

CCC interaction maximizes the combination of both flexible-synergy and market-like effects, and creates comparative advantages that cannot be effectively generated in other types of intra- and interfirm interaction.

3.   Logic of Continuity

CCC interfirm relations, delicately mixing cooperation and competition, inherently face the danger of losing the proper mix. But a logic of continuity, a set of norms and values, sustains continuous relations. The logic, embedded in the concept of network-oriented interpersonalism, is manifested in flexible cooperation. Interpersonalism, in essence, requires one to behave based on adjusting a self-centered evaluation of a given situation by taking into consideration of the expectations of others. Individuals and companies in a network-orientation, which is rather open, fluid, and flexible, base their self-adjustments on the expectation of others, neither from the perspective of subjugation (hierarchical interpersonalism) nor from that of group conformity (group-oriented interpersonalism). Flexible cooperation is based on the logic of continuity, and individuals within the system try to serve mutual interests and maximize gains by understanding the expectations of others, adjusting oneself in accordance with others' expectations and generating flexible-synergy and market-like effects.

The logic of continuity assumes that flexible cooperation with proper conditions can generate many more benefits than operating simply in an open market or inside a company. Its basic principles of interfirm relations are the following: (1) to maintain a sense of trust and cooperation; (2) to mutually share information and capabilities for survival; (3) to meet severe and continuous demands for each other, including effectively responding to competition-generating measures and enhancing competitive advantages in a final product market; and (4) to develop future-oriented and risk-sharing activities.

During an initial testing period that involves a limited number of orders, intensive mutual evaluation takes place. Once interacting partners come to perceive that their mutual interaction has a potential for generating interfirm flexible-synergy and market-like effects at present and/or in the near future, their pattern of interaction changes. They gradually increase the areas of cooperation, interdependence, and sharing. Both sides come to expect continuity in the relationship. As a condition of maintaining the continuity, both sides openly make demands and expect them to be met interactively. This process involves not only the increased sharing of resources, but also more frequent monitoring of whether partners are satisfying needs and demands. Ultimately, to maintain future cooperative relationships, companies have to jointly engage in future-oriented risk-taking activities. Such development strengthens continued CCC relations and generates more interfirm flexible-synergy and market-like effects.

Hence, the logic of continuity, in its concrete form, sets up four criteria of mutual evaluation: (1) whether CCC interfirm relations are generating sufficient benefits at present or are laying a ground for future benefits; (2) whether partner companies are meeting necessary criteria and demands for surviving in a competitive product market; (3) whether partners are effectively responding to diverse competition-generating measures; and (4) whether partner companies are risking themselves in setting up a future path of CCC interfirm relations. These criteria become the basis of evaluation, influencing the selection of cooperation-promoting and competition-generating measures, and differentiating flexible-synergy and market-like effects by partners. The logic of continuity has the underlying values and norms to sustain CCC interfirm relations.

4.   Power Relationships and Logic of Continuity

Power relations further complicate Japanese interfirm relations. Semiconductor manufacturers establish long-term relations with both large-sized and small- and medium-sized companies. But differences in their capabilities greatly influence power relationships, and even differentiate the focus of the logic of continuity.

As Table 1.2 shows, large-sized companies vis-á-vis semiconductor manufacturers tend to have equal or sometimes superior capabilities in technology, management, R&D, and marketing. Resources that both sides have are mutually attractive to share, if a proper cooperative arrangement can be established. Hence, long-term relations between these two partners tend to be power-symmetric with equal bargaining power on both sides. Such relations are characteristic of horizontal interaction. In contrast, small- and medium-sized companies may have sufficient capabilities to satisfy the demands and needs of semiconductor manufacturers. But in general, their capabilities in technology, management, R&D, and marketing are far inferior to semiconductor manufacturers. Consequently, the weaker bargaining power of these companies causes power-asymmetry vis-á-vis semiconductor manufacturers. This is characteristic of vertical interaction.

To students of Japanese studies, horizontal and vertical interfirm relations may suggest horizontal and vertical *keiretsu*. Some semiconductor manufacturers actually develop strong cooperative relations and form *keiretsu*. Cooperative interfirm relations in this book, however, refer to any long-term

Table 1.2 Difference Between Vertical and Horizontal Interaction in Flexible-cooperation-based Long-term Interfirm Relations

INTERACTION		CHARACTERISTICS OF PARTNERS VIS-á-VIS SEMICONDUCTOR COMPANIES	POWER RELATIONSHIP	EFFECTIVENESS OF COOPERATION-PROMOTING MEASURES	EFFECTIVENESS OF COMPETITION-GENERATING MEASURES	LOGIC OF CONTINUITY
Flexible-cooperation-based Long-term Interfirm Relations (Competitive-*cum*-cooperative Interfirm Relations)	Horizontal Interaction	-Large-sized companies -Equal capability -Equal bargaining power	Symmetric	Medium	Medium +	Performance-oriented (emphasis on performances in technology, quality, cost, delivery, and services)
	Vertical Interaction	-Small-and medium-sized companies -Weak in capability -Weak bargaining power	Asymmetric	Medium+	Medium	Human-relations-oriented (emphasis on cooperation, trust, and human relations)

relation that is perceived to involve some cooperation, and this study does not restrict findings to *keiretsu* relations.

Both vertical and horizontal interaction is based on flexible cooperation, but cooperating partners in vertical and horizontal relations respond differently to cooperation-promoting and competition-generating measures, hence showing different characteristics in CCC interaction. Small- and medium-sized companies, or vertically-related companies, need help in technology, production, management, training, and finance, indeed, in almost every area. For them, any kind of help is highly welcomed, especially if it is done with a strong offer of cooperation from large-sized companies. Since cooperation is a manifestation of the logic of continuity, giving such help to vertically-related companies is an expression of mutual trust and their willingness to continue working for the future, but with very high semiconductor companies' expectation that vertically related companies meet required standards and satisfy demands. However in some cases, what is perceived to be cooperation at the very beginning may turn into highly exploitative behavior, which asymmetrical power relations can easily generate. If this occurs, smaller companies may try to exit, or may persevere if they cannot find any alternative. But such behavior is less likely to occur if both companies are basing their behavior on the pursuit of mutual gains.

In return for help provided by semiconductor manufacturers, vertically-related companies can offer a strong sense of cooperation in responding to the needs and demands of large-sized companies, demands which horizontally-related companies may not accept. Both sides try to mutually maintain a strong sense of favorable cooperation. Hence, vertical interactions tend to show a higher level of response to cooperation-promoting measures than competition-generating ones (see Table 1.3).

What sustains such a strong sense of cooperation are human-relations-oriented norms and values. Sometimes, the intensive assistance offered by semiconductor manufacturers and the difficult demands that vertically-related companies face are not understandable in regular market interactions. Prices for these types of works would be enormous if contracted in the market. Though such incidents may not be frequent, managers understand that they can happen at any time. Maintaining such relations becomes highly valuable and very cost effective when done with a strong sense of cooperation. Besides, many minor adjustments that occur regularly can be performed more smoothly and quickly with very minor costs. Hence, even asymmetric power relations can lead to enormous mutual gains when cooperative relations are well established. Companies involved in these relationships will achieve a type of gain only workable under power asymmetry. In this way,

Table 1.3 Flexible-synergy and Market-like Effects by Type of Governance

GOVERNANCE		FLEXIBLE-SYNERGY EFFECTS	MARKET-LIKE EFFECTS
Hierarchical (Intrafirm Relations)		Medium	Low
Competitive-*cum*-cooperative	Horizontal Interaction	Medium +	High
(Flexible-cooperation-based Long-term Interfirm Relations)	Vertical Interaction	High	Medium +
Market		Low	Medium

the appeal of maintaining good and strong human relations has also an economic base. Efforts to maintain such relations result in a more favorable response on both sides to cooperation-promoting measures than to competition-generating ones. In other words, vertical interfirm relations develop the human-relations-oriented logic of continuity.

In contrast, large-sized partners, or horizontally-related companies, are as capable as the semiconductor manufacturers. They cooperate when it is clearly necessary. They are willing to engage in practices like sharing technological and market information, developing technology jointly, engaging in quality assurance programs and value engineering and analysis, coordinating operations, delivering just-in-time, and providing extra services. Owing to their high capability, they respond very effectively to evaluation, monitoring, and other competition-generating measures. They also operate efficiently in the market-like environment created by semiconductor manufacturers, while they also like to promote further cooperation and maintain long-term relations. But horizontal interfirm relations do not require the spread of cooperation to all aspects of the relationship. Thus, the sense of cooperation is weaker, and large-sized companies respond more effectively to competition-generating measures than cooperation-promoting ones (see Table 1.3).

Companies in the market are also required to respond effectively to competition-generating measures and achieve satisfactory performances in technology development, quality, cost reduction, delivery, and services. What, then, differentiates their responses between horizontally-related companies and those in the market? The difference lies in whether companies have a strong adherence to the logic of continuity, namely, the continuity of cooperative relations, mutual sharing of information and capabilities, mutual fulfillment of expectations, and risk-taking activities. Once long-term cooperative relations are involved, the foci of evaluation shift from superficial aspects of performance to the efforts and attempts to fulfill the logic of continuity and their actual outcome. I shall call this logic the performance-oriented logic of continuity.

5.   Cooperation- and Competition-oriented CCC Interaction (Influence of Market Conditions and Functions)

Flexible cooperation can be achieved as an outcome of transformation from either traditional-and-rigid-cooperation-based interfirm relations or from market interaction. As Figure 1.2 shows, to make traditional and rigid interfirm relations more flexible and dynamic, companies introduced competition-generating measures to their cooperating partners, monitored and examined partner's compliance with those measures, and adjusted the intensity of their cooperation based on evaluations. Cooperation, however, remained as the basic orientation of interactive behavior, while competition-generating measures became a means for adjusting and sustaining continuous coop-

Figure 1.2 Market Competition and Transformation to Long-term and Pseudo-long-term Flexible Cooperation, and Cooperation- and Competition-oriented CCC Interfirm Rerations

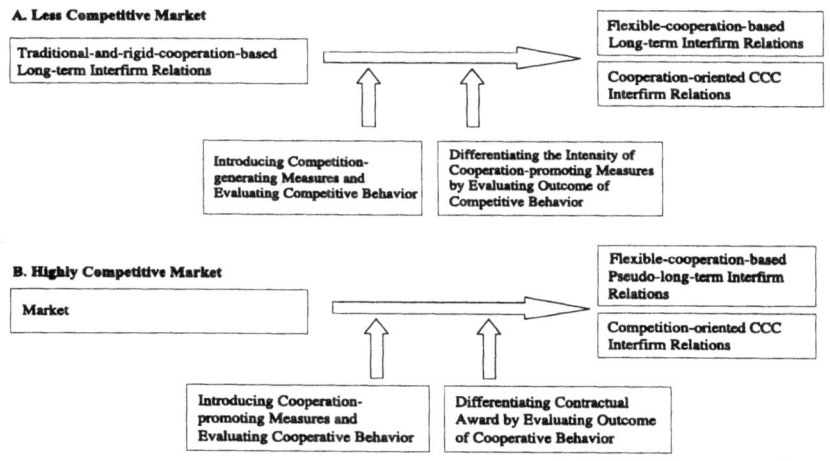

erative relations. There were implications of uncertainty, threat, and instability in cooperative relations (Okada, 1993). I call interaction based on this type of flexible cooperation cooperation-oriented CCC interaction.

In contrast, companies can develop flexible cooperation from market interactions by introducing cooperation-promoting measures, monitoring and examining partner's compliance with these measures, and differentiating contractual awards based on evaluations. In essence, cooperation simply remains as a means of reducing the threat and uncertainty of what are basically very unstable relationships. Cooperation then, results in pseudo-long-term relations, in which companies win contracts consecutively and give the appearance of long-term relations. In reality, there is a little commitment to the relations, though cooperative behavior increases the probability of winning contracts consecutively. This type of interaction – competition-oriented CCC interaction – is closer to market transactions than cooperation-oriented CCC interaction, although cooperative behavior still generates additional comparative advantages.

For some scholars who perceive long-term relations as something much firmer, competition-oriented CCC interaction may not be considered to be a form of long-term relations. Such a form of interaction became possible only under the environment of flexible cooperation, which heavily weighs mutual interests and gains among interacting partners. The logic of continuity is still firmly followed. As a means of winning competition, companies mutually share information and capabilities for survival, meet rigorous expectations, and try to establish a sense of trust and cooperation. Once a sense of trust and cooperation is established, the probability of winning the next contract is increased. Within such a process, attempts to keep winning a sense of trust and cooperation become a very important key for continued relations, especially for developing future-oriented and risk-taking activities. The major difference from cooperation-oriented CCC interaction is that the lack of guaranteed future relations introduces continued rigor and pressure for obtaining trust and cooperation.

A company's choice of CCC interaction also differs according to the conditions of a product market, the relevant features, functional areas, and the characteristics of partners. Cooperation-oriented CCC interaction is predominantly found in a semiconductor manufacturer's stable long-term relationships with parts, material, and equipment suppliers, whose markets tend to be less competitive than the semiconductor market. Producing semiconductors requires precision and a very tight control over quality and environment. When the cropping rate of chips from a silicon wafer declines drastically, an immediate search for its causes starts. The company needs to test extensive areas of the factory environment such as dust in the air, the quality of parts and materials such as the silicon wafer and even the water, and the precision of machines. Some crucial materials such as the silicon wafer are numbered and computer-controlled so that the producer and the lot number are immediately identifiable. Prompt corrective actions become necessary on the part of suppliers. If the problem is machinery, then immediate repair work becomes necessary by the manufacturers. In some cases, equipment manufacturers station their engineers in customers' factories to be ready for prompt repair work. Given the nature of supplies, interfirm relations become highly cooperation-oriented, and the market tends to be less competitive. For this type of operation, vertically-related companies are easier for semiconductor manufacturers to work with, since they are highly cooperative. Sometimes semiconductor manufacturers try to nurture the capabilities of small companies, so that cooperative working relations can be developed. But if this is not possible, then semiconductor manufacturers tend to work with technologically-sophisticated horizontally-related companies. This is often the case with semiconductor manufacturing equipment suppliers.

Competition-oriented CCC interaction is found in semiconductor manufacturers' relations to buyers. Given severe market competition in sales, the former try to turn highly unstable buyer relations into pseudo-long-term relations by successfully developing means of cooperation. In this type of interaction, to prevail competitively is the predominant motive of the sellers, and they promote cooperation as a means of obtaining buyers' information more quickly than competitors, securing relationships with them, and reducing uncertainties in sales activities. Such a highly competitive situation allows buyers to realize benefits from both market competition and cooperative behavior, and even to use the level of cooperation as one criterion for stimulating competition.

This tendency is especially strong in horizontal relations between semiconductor manufacturers and large-sized customers. Since the latter has many projects for new products that require forefront technology, it becomes essential for the former to be involved in planning, product development, designing, and engineering stages. To become a project partner in those stages provides diverse benefits of access into forefront technology, the potential to supply a large quantity of parts for a long time if the product becomes a good hit, and the potential to join other new projects. Such processes require semiconductor companies to develop good cooperative relations with personnel in the planning, product development, designing, and engineering sections of customer companies, to compete against other suppliers by partially revealing segments of crucial technological information in order to show technological superiority, and to suggest new ideas for selling their products. Such behavior also benefits buyers greatly in developing a new product and encourages them to contact several other highly competent companies to obtain better and newer insights. Consequently, competition for developing cooperative relations becomes very keen. Success is based on whether a semiconductor manufacturer can become a good cooperating partner and demonstrate superior technology and concepts while preventing the leakage of crucial technological information. Thus, the conditions of a product market and functional areas and the characteristics of partners differentiate semiconductor companies' choice of CCC interaction.

6.   Measures Generating Flexible-synergy and Market-like Effects

Thus far, I have discussed flexible-synergy and market-like effects in abstract terms. It is quite important to introduce the concrete measures that generate these effects. There are four conditions that generate flexible-synergy effects: (1) the need for mutual assistance; (2) the need to acquire information; (3) resource interdependence; and (4) the need to stimulate innovation (see Table 1.4). In the logic of continuity, the first three needs lead to the mutual sharing of information and capabilities for survival and the last one to future-oriented and risk-taking activities. Stimulating innovation is treated separately from the other measures of future-oriented risk-taking activities, since technological innovation is the most vital element for future survival. Included in mutual assistance are some of those activities not related to innovation stimulation, such as (1) risk-sharing by forming a joint venture or franchise contract, (2) help in crisis situations through procurement, finance, and personnel exchange, (3) joint establishment of facilities, (4) production alliances, and (5) coordinated investment. Some measures may overlap in categories.

a.   *Measures Generating Flexible-synergy Effects*

(1)   Measures for Mutual Assistance
Mutual assistance is the way interacting companies help each other achieve shared goals. In post-World War II Japan, these practices were established to stabilize companies' performance in the highly unstable economy, though some methods were inherited from the war-time practices. Some of the reasons that Japanese companies developed interfirm cooperation are related to historical conditions. The primary factor was the severe labor disputes in the late 1940s and the early 1950s, which compelled management to accept permanent employment and the seniority-pay system (Cole, 1979; Dyer, et al.,

Table 1.4 Measures Generating Flexible-synergy and Market-like Effects

---

A. Flexible-synergy Effects

---

1. Measures for Mutual Assistance

-help in crisis situations through procurement, finance, and personnel exchange
-small size and cooperative production to overcome market fluctuation
-small size and cooperative production to maintain permanent employment and harmonious labor-management relations
-risk sharing by forming a joint venture or franchise contracts
-joint establishment of facilities (especially a testing center)
-computer coordinated market information exchange and production integration
-stable mutual ownership
-coordinated long-term planning

-financial guarantee and discounting
-financing with favorable terms
-technology transfer
-sharing complementary assets
-providing training
-cooperation to maintain high quality of parts and materials
-alliances on production and sales
-coordinated automation, robotization, computer control and just-in-time production
-production and marketing schedule coordination
-establishing mutually acceptable standards
-investment coordination
-strategy coordination
-coordination in quality circle, total quality circle, and total productivity maintenance activities

---

2. Measures for Information Acquisition

-exchange of information on product quality
-exchange of market actor credibility information
-exchange of information on opportunistic behavior from networks based on trust
-exchange of upstream market and technology information

-developing information networks to effectively and quickly raise the awareness and level of technological capability
-developing information exchange partners over production, market, and technology

---

3. Measures for Resource Interdependence

-to maintain specialization for production cooperation
-personnel dispatching
-cooperation on sales and procurement

-just-in-time coordination
-complementary technology interdependence
-cooperative procurement for resource security

---

4. Measures for Stimulating Innovation

-design and production coordination
-simultaneous engineering
-R&D cost sharing and prevention of duplicate effort
-quality circle and total quality circle coordination
-value analysis and engineering with bonuses to outside firms

-quick innovation to catch up with shorter product life cycle
-joint R&D
-cross licensing

B. Market-like Effects

---

1. Measures for Market-Linked Adjustments

-constant reduction of cooperating firms
-explicit penalty for late delivery and poor quality
-severance of old ties and entry of new innovative firms
-nurturing multiple vendors for a product
-promotion of diversification and financial independence

-parts pricing method
-tough negotiations (constant demand for lower prices)
-market-linked price reduction based on agreed schedule
-value analysis and engineering with bonuses to outside firms

---

2. Measures for Reevaluation Adjustments

-supplier evaluation (delivery capability, cost reduction capability, quality standard, services, etc.)

-quality circle system survey
-contract renegotiation for each model (e.g. the automobile industry)

---

3. Measures for Resource Interdependence Adjustments

-volume guarantee especially for small producers
-monthly quantity adjustments

-contractual assurance with built-in flexibility

1987, Smitka, 1991). This guarantee compelled companies to remain as small as possible, since laying off workers resulted in labor disputes. To cope with the increased demand for products with the smallest additional employment, interfirm cooperation became an attractive corporate strategy.

The second factor was the unstable economic conditions in the 1950s. Small size was a necessary part of coping with fluctuations in the economy. Efforts to remain small led to specialization and resulted in a significant amount of subcontracting and interfirm cooperation, continuing the pre-war Japanese tendency of specialization (Cole and Yakushiji, 1984; Ohno, 1982; Fruin, 1992). Interfirm cooperation among smaller companies also enabled Japanese corporations to create a scale of economy in production without additional bulky bureaucratic costs caused by internal production (Yoshitomi, 1990). Furthermore, companies often jointly established facilities such as testing centers (Okada, 1999). Small size, however, increased the risk of takeovers and the influence of stockowners. Hence, interwoven stable ownership was developed as a convenient solution to these problems, enabling companies to implement long-term planning (Futatsugi, 1990; Yoshitomi, 1990).

The third factor was the effort to reduce uncertainty in operations and investments. Needs for securing resources, producing high-quality goods with much lower costs, preventing opportunistic behavior, and sharing the risks of investment made it necessary for a company to join or formulate stable networks of production and information, sometimes linked by ownership and personnel exchange, sometimes purely by a sense of cooperation. When cooperation was well developed, interfirm relations functioned as mutual insurance, with companies helping each other in crisis situations through favorable procurement, financing, and personnel exchange (Dyer, et al., 1987; Jankowski, 1989). Cooperation developed through such interaction also cultivated grounds for additional mutual support and facilitation, positive feedback, accurate communication, information exchange, and trust. To weaker companies, a stronger organization (1) provided financing with favorable terms, guarantees and discounting, (2) transferred more advanced technologies, (3) rented equipment and materials at the initial stage, (4) gave training, and (5) shared complementary assets (Odaka, et al., 1988; Jorde and Teece, 1989).

Mutual assistance in the 1950s and 1960s, given for the purpose of coping with economic difficulties, was based on a sense of traditional-and-rigid cooperation, but those types of measures later continued to function as a part of differentiated reward to generate flexible-synergy effects. In the 1970s and 1980s, facing severe domestic and international competition in product markets, semiconductor companies developed flexible-synergy-based mutual assistance to generate the dynamic transformation of companies and interfirm relations as well as to build new comparative advantages. In interfirm relations, companies came to cooperate intensively in quality circles, total quality circles, process quality control, and total product maintenance. They also began to implement just-in-time production and coordinate automation, production with robots, computer-controlled production, product development, investments, production scheduling, and marketing (Jones, 1985; Smitka, 1991; Shimokawa, 1990). In the later 1980s, alliances among power-symmetric manufacturers also became very popular (Jorde and Teece, 1989; Contractor and Lorange, 1988).

(2)   Measures for Information Acquisition

Quick access to accurate and valuable information not only functions to help companies maintain a position, but also provides advantages over those who have slower and less access. Valuable information, however, is usually only transmitted between closely and frequently interacting companies, making it necessary to develop favorable interfirm relations (Oaumann and Knoke, 1989; Calder, 1988). Such relations can provide companies with valuable technological and product-market information, stimulate innovative activities, and raise awareness of the business environment with a far lower cost (Lynn, 1984). Vital information about other companies is also transmitted. Reputation about product quality, credibility, or opportunistic behavior spreads with unusually high speed, and companies can greatly benefit especially by preventing disastrous consequences. The importance of such practices remains steady from the 1950s to the present, and they have become an important part of differentiating cooperative reward.

(3)   Measures for Resource Interdependence

Cooperative relationships developed in the 1950s and 1960s extended even to personnel exchange and cooperative bulk purchases. Companies could obtain specialized skills not available within their own organization, secure necessary natural resources, and cut costs through bulk purchasing. Such practices generally diminished in importance by the 1970s, since Japanese companies diversified and increased the size of their operations. But they are still practiced in some vertical *keiretsu* groups. More recently, the development of just-in-time production has intensified resource interdependence between suppliers and buyers.

(4)   Measures for Stimulating Innovation

Interfirm cooperation for technology transfer and innovative imitation played very important roles in developing the basic technological capability of Japanese companies, and these forms of cooperation were often sponsored by industrial associations in the 1950s as well as government technology research cooperatives from the 1960s onward (Okada, 1999). But more recently, competitive product markets and shorter product life cycles have compelled companies to reduce R&D time for innovation (Ohmae, 1985). Interfirm cooperation for technology became very important in the 1980s, since it is a convenient means for obtaining quick innovations through joint R&D and cross licensing and for sharing R&D costs and risks with other companies while preventing duplicate efforts (Jorde and Teece, 1989). Intensified cooperation and coordination through quality circle and total quality circle activities, design and production, value engineering and analysis, and simultaneous engineering strengthened innovative capability (Shimokawa, 1990).

b.   *Measures Generating Market-like Effects*

Flexible-cooperation-based synergy effects alone are inadequate for explaining the dynamics and flexibility of CCC interfirm governance. How well and what type of competition-generating measures cooperating partners can implement makes a drastic difference in their dynamics and flexibility. Competition-generating measures are for (1) market-linked adjustments, (2) re-evaluation adjustments, and (3) resource interdependence adjustments (see Table 1.4).

(1)   *Measures for Market-linked Adjustments*
These measures enable companies to change the characteristics of interfirm relations corresponding to market movements. Although a sense of commitment and reciprocity is quite important especially in vertical CCC interaction, control and domination or subjugation and complete dependence are not considered a wise interfirm relation. As is well known in the automobile industry, an automobile assembler often promotes suppliers' increased independence by recommending the diversification of products and customers and promoting financial independence (Odaka, et al., 1988). The underlying rationale for this behavior is that the more diversified its customers are, the less dependent the small company is on the major customer, thus creating a more stable operation with a wider range of received information. This does not mean a cooperative relation is terminated. As a matter of fact, it is understood that promoting the independence of suppliers brings more benefits to an assembler as long as a long-term relationship is well maintained. Hence, suppliers' experiences in a competitive product market also contribute to the improvement of a stronger company's operation.

Despite well-developed long-term cooperation, some accidents like late delivery, poor quality, high price, and poor services do occur. To prevent such accidents and even to stimulate competition among suppliers (the multiple vendor system), a buyer evaluates and grades a supplier. The buyer also transmits messages about problems and identifies a supplier's position in terms of maintaining long-term or pseudo-long-term relations, sometimes with an explicit penalty. Repeated accidents result in severing ties, although a probation period is usually given.

Furthermore, cooperative relations sometimes involve strategic pricing based on a part-pricing method or on long-term price-reduction planning (Smitka, 1991; Asanuma, 1985). A part-pricing method makes it mutually obligatory to reveal the prices of detailed components, and to plan a long-term price reduction of designated parts. Such component-cost adjustments look beyond the cost of parts and materials, and even require the itemization of costs for every aspect of operation such as administration, financing, and R&D. This method compels less dynamic companies to improve operations and enhance future competitiveness. Market-linked adjustments, however, are not always for tightening operations. When product-market conditions aggravate, companies are also allowed to raise prices for a fixed duration, disregarding contracts.

(2)   Measures for Reevaluation Adjustments
Strict and constant evaluation over long-term suppliers' performances allows companies to propose performance improvements and adjustments, and generate reevaluation adjustments. Supplier evaluation is made on characteristics such as delivery, cost reduction, quality standards, services, and fundamental operational capability. Quality circle surveys are also conducted to evaluate suppliers' managerial capabilities of implementing quality circle activities, which are considered the basic components of production dynamics. These evaluations trigger new and tougher negotiations and give suppliers some pressure for future improvements.

(3)   Measures for Resource Interdependence Adjustments
Adjustments for resource interdependence can take two directions: one is to contractually guarantee the amount of purchases, while the other is to allow flexibility in actual quantity delivery. The combination of these contradictory elements can be identified as contractual assurance with built-in flexibility. What actually happens is that when the demand of a buyer's final product is slack, a buyer can request, despite the contractual specifications, a delayed delivery of parts or the delivery of a reduced amount to adjust to a reduced production schedule. Conversely, when the demand increases, a buyer can request an early delivery and make additional purchases. The

degree of adjustment may differ by the capability of the supplier. Weaker suppliers are usually spared from frequent fluctuations in their actual delivery, while stronger suppliers face more monthly adjustments with purchasers.

Therefore, such competition-generating measures as pricing and cost evaluation, penalty for failing to comply with demand, constant evaluation, and re-negotiation and selection create artificially a market-like environment in cooperation-oriented CCC interactions. In contrast, severe competition in a product market allows buyers to easily implement these measures in competition-oriented CCC interactions. Whether CCC interaction is cooperation- or competition-oriented, without continuous cooperation it is not possible for partners to implement such adjustment measures.

E.   Propositions for Empirical Inquiry

The efficiency of companies' operations is dependent on having the right partners for given contingencies, interacting with partners based on a set of practices and measures, and harmonizing interactions under the right governance. Furthermore, governance has to fit to a company's contingencies in each functional area in order to generate efficiency. Flexible cooperation, a manifested behavior of network-oriented interpersonalism, is a very important part of Japanese companies' contingencies. Flexible cooperation is more open, flexible, and fluid than traditional cooperation, and accepts cooperative bargaining for maximizing mutual gains rather than moral obligations as the basis for judgment. Despite its bargaining orientation, interactive behavior is still based on interpersonalism, where fulfilling mutual expectations is considered of vital importance, and mutual gains are calculated for the long term, rather than the immediate future.

Given the turbulent corporate environments that exist, long-term mutual gains cannot be maintained simply by promoting cooperation. To generate more flexibility, fluidity, and openness to companies' operations, it became necessary for Japanese companies to implement competition-generating measures while rewarding partners with differentiated cooperation. Consequently, flexible cooperation came to blend delicately cooperation-promoting and competition-generating measures in interfirm relations, successfully generating flexible-synergy and market-like effects. Such CCC interactions came to have comparative advantages that cannot be generated by the market or hierarchical governance.

Where an environment is conducive to such interactive behavior, a wise decision for top executives is to find a type of partner who will operate well in the environment of flexible cooperation. Hence, as the basic mode of business practices, Japanese companies should have chosen horizontally- and vertically-related companies, engaged in CCC interaction, and acted under the norms and values of CCC interfirm governance, rather than within market or hierarchical governance. CCC interaction in these companies would then show characteristics including medium levels of cooperation, interdependence, and frequency of interaction, and mutually shared interests and goals. These qualities should be less identifiable in spot interactions, and very present in internalized interactions. CCC interaction should also show a lower level of suppressive structural control than would internalized interaction.

The choice of interaction also generally differs by functional areas. In areas where human interaction is a vital part of developing business relations, such as procurement and sales activities, CCC should become predominant. In areas where highly asset-specific factors are involved and where corporate information must be kept secret, such as R&D activities, internalized interaction should become predominant. Hence, the core propositions to be examined in Chapter Three are as follows:

Proposition 1: Japanese semiconductor companies rely more heavily on CCC governance, consisting of horizontal and vertical CCC interaction, than on market or hierarchical governance.

Proposition 2: Companies with CCC interaction show medium degrees of cooperation, interdependence, and frequency of interaction, and mutually share interests and goals. Those with spot interaction manifest lower degrees in these characteristics, while those with internalized interaction show high degrees.

Proposition 3: High asset specificity and the need to maintain corporate secrets in R&D compel companies to use internalized interaction, while other functional areas that extensively involve human relations rely on CCC interaction.

The type of governance under which a company operates should differ according to market conditions, business function, and type of partner. When cooperation is required for actions such as the procurement of parts, materials, and equipment and when market conditions are less competitive, the environment is conducive to cooperation-oriented CCC interaction. In these interactions, competition-generating measures should be implemented as a means of making cooperation more flexible. But when market competition is very severe and cooperative relations can help a company to win competitions, the environment becomes conducive to competition-oriented CCC interaction. In competition-oriented CCC interactions, cooperation becomes a means for reducing uncertainty and winning competition. These conditions are more often found in sales activities. Hence, another proposition to be examined in Chapter Three is as follows:

Proposition 4: Procurement relationships tend to develop cooperation-oriented CCC interactions, while sales relationships tend to develop competition-oriented procedures.

The comparative advantage of CCC interaction is its capability for generating higher degrees of flexible-synergy and market-like effects than other types of interactions. Internalized interaction generates flexible-synergy effects, but its effectiveness is limited because it disallows competition-generating measures. In contrast, spot interaction generates market-like effects, but these interactions are usually one-time oriented, hence incapable of promoting cooperation.

Vertically-related companies should tend to respond more effectively to cooperation-promoting measures than horizontally-related companies do, because of their emphasis on a human-relations-oriented logic of continuity. Horizontally-related companies should respond more to competition-generating measures, owing to their emphasis on a performance-oriented logic of continuity. Hence, the other core propositions examined in Chapter Four and Chapter Five are as follows:

Proposition 5: Companies with CCC interactions can generate higher degrees of flexible-synergy and market-like effects than those with spot and internalized interactions.

Proposition 6: Vertically-related companies generate more flexible-synergy effects than horizontally-related ones, while the latter generate more market-like effects than the former.

Proposition 7: Vertically-related companies behave based on a human-relations-oriented logic of continuity, while horizontally-related ones act based on a performance-oriented logic of continuity.

Chapter Two

Technological Development, Corporate Strategies and Market Competition

Efficiency in each company's internal operations is the primary source for explaining the Japanese semiconductor industry development. But competitive-cum-cooperative (CCC) interfirm relations add incomparable advantages to the efficiency and flexibility of semiconductor companies' operations. One important factor influencing a semiconductor company's choice of CCC interfirm relations is the condition of the product market. There are three aspects of the market to consider. First, severe competition in the semiconductor market has significantly contributed to the dynamic development of the industry. Secondly, milder competition in parts, materials, and semiconductor equipment markets provided a favorable environment for developing interfirm cooperation and CCC interfirm relations. And finally, cooperative networks and relations led to some radical innovations, even if they did not prompt major breakthroughs. Innovation occurs when companies have chosen corporate strategies suitable to each stage of their development and when they accumulate high-level technological capability. The historical relationships among these factors – product market conditions, technological capability, and corporate strategies – have been vital in the development of the Japanese semiconductor industry. As a background to understanding the dynamics of the industry, I shall trace the historical relationships among the three factors.

In each stage of semiconductor industry development, product market competition and company technological capability greatly influenced and restricted corporate strategies and the types of technologies and products that were developed. Under these restrictions, companies had to make decisions about developing technology with or without a foreign license, for consumers or military applications, with intensified interfirm cooperation or severed relations, and in a market-driven or technology-driven way.

Since technological sophistication and corporate strategies are closely interrelated, I shall use these two factors to develop a typology of technological innovativeness and to identify the stages of technological development in the Japanese semiconductor industry. Since the development of the industry parallels advancement in memory technology, I shall focus mostly on memory, which transformed from the transistor to the integrated circuit (IC) to the large-scale integrated circuit (LSI) to the very-large-scale integrated circuit (VLSI) and finally to the ultra-large-scale integrated circuit (ULSI) by 1992. For each kind of memory, I shall trace the history of interrelations among technological development, corporate strategies, and market competition (Okada, 1989b; Okada, et al., 1994).

Methodologically concerned, I have identified major technological developments chronologically and made rough-estimates of the time lags between U.S. and Japanese innovations in selected semiconductor products (Appendix Table 2.2). This table, along with materials provided by top-level managers of semiconductor companies and literature survey, gave me insights on which stage of innovativeness Japanese companies were in. The corporate strategies of two companies were identified from both interviews and the literature survey.[1] Changes in corporate strategy usually appear explicitly in organizational goals and structures (Appendix Tables 2.4 and 2.5). Two yearbooks have detailed records of the company's entry into the market, new technologies, technical agreements, government policies, and incidents in the semiconductor industry. Along with other literature, the yearbooks became the source to estimate the intensity of market competition (Denpa Shinbunsha 1962 - 1997; Press Journal, 1985 - 1996) (Appendix Table 2.3). These three sets of information became the base of this chapter.

A.   Typology of Technological Innovativeness

Schumpeter argues that a state-of-the-art breakthrough in technology brings a drastic change to the existing technological system and replaces an old industry with a new one (Scherer, 1986). The invention of the germanium point contact transistor by J. Bardeen and W.H. Brattain of the Bell Telephone Laboratory in 1947 was surely such a case. Their invention made a revolutionary change in the electrical industry, displacing vacuum tubes, miniaturizing electrical goods, and even giving birth to the semiconductor industry.

Semiconductor technology advances quite rapidly. For example, in 1990

---

[1] The managers I contacted played crucial roles in the development of semiconductor technology and divisions. The number of interviewees was three in Company A and six in Company B.

Hitachi announced the world's first 64M-bit dynamic random access memory (DRAM) IC, which could contain 140 million circuit elements in about 185 mm with a minimum line width of between 0.3 and 0.4 micron (Nihon Denshi Kikai Kogyo Kai, 1991:41). When it started producing its samples in 1993, NEC, Toshiba, and Hitachi announced their success in developing the world's first 256M-bit DRAM IC (line width of 0.25 microns for NEC and Hitachi and 0.4 microns for Toshiba). And only two years later, NEC and Hitachi succeeded in the experimental production of the 1 giga-bit DRAM (line width of about 0.14 microns), which can contain 400 pages of newspaper (Press Journal, 1994:277, 367; 1996). Although there have been no state-of-the-art breakthroughs in this field recently, technology has been advancing rapidly.

Semiconductors are used in a wide range of products. The sales of transistors and other semiconductors in Japan started at ¥6.1 billion in 1957 and grew to ¥179.6 billion in 1970, when IC sales began at ¥53.3 billion. IC and semiconductor sales combined grew to ¥864 billion in 1980, to ¥3.62 trillion in 1990, and to ¥4.79 trillion in 1995 (Denpa Shinbunsha, 1962-1982; Okada, 1989b; Nihon Denshi Kikai Kogyo Kai, 1991:109; Denpa Shinbunsha, 1997:40-41). The sales volume expanded almost eight hundred times between 1957 and 1995. These figures clearly suggest that the invention by Bardeen and Brattain made a revolutionary change in life styles, and developed a new industry, involving highly advanced, quickly changing and costly technologies.

Japanese semiconductor contributions, however, were mostly through incremental innovations, though some of the more recent ones were radical. The early development of the Japanese semiconductor industry followed a pattern of transferring and imitating foreign technologies as well as innovating beyond the best foreign technologies. What made Japanese development unique and strong was its capability to adopt, adapt, and indigenize foreign technology (Inukai, 1981:92), advancing technological development from pure imitation to innovative imitation and then to creative innovation (Crawford, 1983). In a sense, innovation occurred through the accumulation of constant improvements and minor technological developments (Imai, M., 1986). Although each innovation may have been minor, the cumulative and continuous process enabled Japanese companies to advance their technological capabilities and innovativeness to a point where some radical innovations started taking place. And in each stage of development, product market competition and corporate strategies played indispensable roles in determining the content and targets of innovation.

When companies had to cope with uncertainty in a product market experiencing recessions or severe market competition, one of the most important ways to cope with problems was to enhance technological advantages. This sometimes required drastic changes in corporate strategy and management. Whether a company succeeded in its transformation was very much determined by strategies formulated by top decision-makers and their management effectiveness. Companies that successfully developed technology have often had highly devoted and self-sacrificing researchers and engineers who were genuinely motivated by an interest in technological developments, rather than an interest in profit making. They tried to develop high-risk and future-oriented research programs (Riggs, 1983), though they often faced the discouragement of top executives. Examples of poor decision-making and mismanagement are also abundant in this highly competitive product market.

Hence, the corporate strategies used in each stage of technological development made crucial differences in future technological capabilities and performances. Top executives are responsible for many decisions including the amount of R&D expenditure, the indigenous development of technology vs. dependence on foreign technological licenses, monopolization of technology vs. cross-licensing of technology, market-driven vs. technology-driven innovations, R&D for commercialization vs. technological application, R&D for technological application vs. basic technology, etc.

Figure 2.1 shows the degrees of innovativeness, reflected in technological development and corporate strategies, in the development of Japanese semiconductor industry.[2] There are two periods in the stage of imitation: experimental imitation and imitation for production. In the first period, a company engages in an experiment simply to reproduce an already existing foreign product for the purpose of accumulating basic information and know-how in a laboratory with little R&D cost. In this stage a clear vision of corporate strategy does not exist, but what makes a difference is the researchers' insight about the future of experimental technology and their own devotion to its development.

The second period, imitation for production, requires more sophisticated know-how and additional major technologies for commercial production. These assets are mostly obtained by contracting technical agreements with foreign companies. This stage, however, requires additional R&D and a

---

[2] I have revised Crawford's (1983:98-105) categorization to be more suitable to the Japanese semiconductor industry. In order to indicate the degree of innovativeness, Crawford used the three stages of "imitative/emulative," "adaptive," and "pioneering uniqueness," rather than "imitation," "innovative imitation," and "creative imitation." Also, his classification of "quick second," "segment franchise," and "economic low price" in the "imitative/emulative" stage was not considered important for this research.

commitment to mastering production technology, which involves highly critical corporate decisions regarding the type of technology to develop and potential partners for technical agreements.   Immediate commercialization is strongly emphasized, partly because a large sum of money is required for patent royalties and technical fees and partly because any company with a basic capability can quickly enter a product market with foreign licensing. The need to reduce the heavy burden of patent and technical-fee payments and to achieve product differentiation from competitors equipped with the

Figure 2.1 Degrees of Innovativeness: Indicators and Management Strategies

	Imitation	Experimental Imitation	Accumulation of basic information and knowledge
			Duplication of what was produced previously
			Lack of commercial production capability
			Little R&D expenditure
			Foreign technology dominant
		Imitation for Production	Commercialization of products
			Small R&D expenditures for commercialization
			Foreign technology dominant
Innova-tiveness	Innovative Imitation		Changing technology more suitable to the market
			Medium R&D expenditure for commercialization
			Market-driven innovation
			Small accumulation of indigenous technology
			Foreign technology is more influential than domestic technology
	Creative Innovation	Application Engineering	Application of knowledge to new areas
			Development of new technology via extension of knowledge
			Larger R&D expenditure for commercialization than applied field
			Technology-driven innovation
			Some accumulation of indigenous technology
			Domestic technology is more influential than foreign technology
		Leveraged Creativity	Utilizing knowledge and skills to create drastically new products
			Marriage of marketing and technical inventiveness
			Larger R&D expenditure for applied fields than commercialization
			Technology-driven innovation
			Significant degree of indigenous technology accumulation
			Domestic technology is more influential than foreign technology
		State-of-the-art Break-through	Most drastic changes in the technological system
			Inventing technology with the possibility of forming new businesses and industries
			Extra-large R&D expenditure for basic science
			Technology-driven innovation
			Indigenous technology dominant
			Domestic technology dominant

Note: Terms used by Crawford have been somewhat modified.
Sources: C. Merle Crawford, New products management, (Homewood, Ill.: Richard D. Irwin, Inc., 1983), 100.
           Ryuzo Sato, *Gijutsu no keizaigaku* (Economics of technology), (Tokyo: PHP Kenkyusho, 1985).

same licensed technology force a company to gradually increase its R&D budget for the commercialization of products (Sato, 1985).

The next stage is innovative imitation. This term means the alteration of a foreign technology in some way to make it more suitable to a local product market (a market-driven innovation). Partly because of some accumulation of internal technologies (possessed by a company) and domestic technologies (possessed by other Japanese companies) and partly because of increased product-market competition with new entries, each company engages in more product differentiation and minor competition-driven (market-driven) innovations. Such a strategy forces a company to increase R&D expenditure for commercialization and allows it to accumulate technologies internally, although foreign technology is still dominant in this stage.

The most advanced stage is creative innovation, in which a company develops an innovative product and engages in aggressive marketing for the product. Once internal and domestic technology accumulates sufficiently and becomes more dominant than foreign technology, under severe market competition, a company shifts the nature of corporate strategy from market-driven innovation to a technology-driven one (Crawford, 1983). Since minor product differentiation is no longer sufficient to maintain competitiveness in a product market, major product differentiation becomes essential.

Even in the creative innovation stage, there are differences. The higher the level of creative innovation a company pursues, the more innovative and technology-driven it needs to be and the more R&D expenditure it has to make. Companies without large R&D budgets can do only application engineering or the routine application of engineering knowledge to an area in which it has not been previously applied. If new technology develops, that innovation is but an extension of existing knowledge, and the target of innovation is still commercialization rather than technological development in an applied field.

With a larger commitment to R&D, companies can engage in leveraged creativity, a process in which they utilize knowledge and skills already available to create new configurations, features, forms, and dimensions. The central elements are creativity in the marketing department and in the laboratory, with the latter providing a physical form for any creative marketing insight and the entire process yielding a marriage of marketing and technical innovation. This requires a high level of indigenous technological accumulation. Leveraged creativity, in contrast to application engineering, is more concentrated on technological development in applied fields, targeting future commercial applications.

Creative innovation can also be expressed in a state-of-the-art breakthrough. This kind of innovation brings the most drastic changes in the existing technological system and even creates opportunities for new businesses and industries. In this pattern, leadership in the industry as well as abundant financial resources allow a company to search for a state-of-the-art breakthrough to further strengthen its leadership in product markets. The needs to maintain industry leadership compel top decision-makers to engage in basic research and very high-risk innovative projects, requiring an unprecedented amount of R&D expenditure (Sato, 1985).

Japanese strategies have been directed increasingly toward reliance on internally and domestically accumulated technology, and toward advancement from commercialization to market-driven and to technology-driven innovation. Within technology-driven innovation, emphasis also has shifted from commercialization to the applied field and, finally, to basic research.

B.   Development of the Transistor in Japan

1.   Experimental Imitation Period (1948 - 1953)

The Japanese semiconductor industry originated in 1948, when a study group on the transistor was informally established in the Electro-Technical Laboratory (ETL, Denkishikenjo) of the Ministry of Commerce and Industry (currently Denshi Gijutsu Sogo Kenkyusho of the Ministry of International Trade and Industry [MITI]). The group's activities expanded when the government officially subsidized its activities in 1949. It consisted of researchers from ETL and the Telecommunication Research Institute of the Ministry of Communication (TRI, Denkitsushin Kenkyusho), professors from the Tokyo University and the Tohoku University, and eight researchers from private companies. The group members concentrated their efforts on obtaining and discussing basic information from the library in the headquarters of the Occupation Force and other sources. The library was then the only place in Japan where the Bell System Technical Journal was available to the public (Aida, 1991).

As expected by MITI officials, some of the company researchers in this group played leading roles in influencing the top-level executives of their companies, reformulating corporate strategies and developing transistor technology. In 1949, they had their first look at a point contact transistor produced by Western Electric (Nakagawa, 1985). This was the beginning of the experimental imitation period of the transistor.

Lack of materials, facilities, and proper environment tormented the researchers, and there were many failed experiments. It was 1951 when TRI, which later became a research institute of the Nippon Telegraph and Telephone Public Corporation, succeeded for the first time in experimental production of a point contact transistor. This Japanese success lagged behind U.S. experimentation by roughly four years (Appendix Table 2.2). The basic information, know-how, and skills accumulated in this experimental process significantly accelerated the rate of experimental imitation of other types of transistors. In the following year, TRI successfully produced a PN junction transistor, lagging only about one year behind the U.S.

In 1953 Japanese companies succeeded in the experimental imitation of the PN junction transistor using an alloying method, again lagging only one year behind U.S. technological development. Despite all these research achievements, many companies were aware that accumulated know-how was not sufficient for mass production. They encountered many new and difficult problems, such as low yield rates, low quality of materials, inadequate bonding processes, semiconductor instability, low quality air filters, purity of chemicals, and the need for dust free environments and more refined temperature controls. Acute awareness of difficulties led companies to contract technical and patent agreements with foreign companies.[3]

The time sequences between technical agreements and success in experimental production differed company by company. A study group member in Company A requested his top-level managers to start a research project for developing the transistor. However, since it was the peak period of vacuum tube production, initial permission was denied. Later, responding to the news about revolutionary technology that was constantly flowing in from the U.S., the company's top executives approved the project (Nakagawa, 1985). Soon after commencing the project, Company A contracted agreements with foreign companies, one in 1952 and another in 1954 (Appendix Table 2.4). With the help of foreign technology the organization successfully produced an experimental point contact transistor and a PN junction transistor with an alloying method.

The case of Company C, the largest producer of vacuum tubes at that time, was similar. Researchers had difficulty persuading top executives to pay serious attention to the transistor. While other companies started conducting research in 1951, this company did not do so until it made a technical agreement with RCA in 1952. Even so, its position in the vacuum tube industry prevented it from advancing research to the stage of mass production

---

[3] Hitachi, Toshiba, and Kobekogyo (later acquired by Fujitsu) contracted with RCA in 1952 and Western Electric in 1954; Fuji Electric with Siemens in 1952, SONY with Western Electric in 1953, and NEC with RCA and General Electric in 1958 (see Appendix Table 2.3).

until SONY announced the transistor radio in 1954 (Nakagawa, 1989).

The case of Company B is quite unique. Young researchers there, one a member in the government sponsored study group could not obtain permission for research from top executives. During their lunch breaks, they developed experimental equipment from scraps found in the company's storage area (Aida, 1991). They were the first researchers working in a private company to engage in research and obtain some success, and they did so without official support of the company. Despite very difficult conditions, their enthusiasm and efforts enabled them to successfully make the first Japanese-produced germanium crystal (Nakagawa, 1985). Persuaded by their success, the company started an official research project on transistors in 1950. While other companies were contracting technical agreements with U.S. companies, researchers and engineers at this company were fiercely debating whether they should import foreign technology or further develop indigenous technology to compete against other domestic companies (Nakagawa, 1985). The initial decision was the latter. However, because they lacked sufficient know-how for mass production, they were unable to continue this course. Consequently, this company made a late start, four or five years behind other companies, but did so with a strong spirit of self-reliance that bore fruit later. Helped by its specialized field of production, the company was not seriously harmed by the delay (Appendix Table 2.5).

The experimental imitation period in the development of Japanese transistor technology was characterized by the constant struggles of forward-thinking researchers to learn about the technology, to persuade company executives of the importance of the technology, and to develop mass-production technology. The lack of technological capability was so extensive that foreign technological dependence was found indispensable.

2.   Imitation-for-production Period (1954 - 1959)

Japanese companies made every effort to catch up with U.S. innovations such as the mesa transistor and the transistor with an alloying diffused method, shortening the time lag to one year for the experimental imitation. Although this period ushered in the very unusual Japanese innovation of the Esaki diode by Nobel Laureate Reona Esaki of SONY, the shortening of the time lag was substantially due to technical agreements with U.S. companies.

Mass production of the transistor did not begin until 1956, four years after the first technical agreements and two years after SONY's announcement of the transistor radio. This delay was due partly to the lack of know-how, and partly to the lack of demand. Many companies still concentrated on products with vacuum tubes. In this market, SONY's bold attempt to produce a transistor radio was quite sensational, especially because it was believed to be indigenously developed and the first in the world. (A U.S. company, Regency, had succeeded in developing the transistor radio six months before the SONY announcement (Nakagawa, 1985)). SONY radio sales accelerated the change from vacuum tubes to transistors.

From the very beginning, Japanese companies were forced to choose different market strategies from U.S. companies. The SONY radio signifies two major characteristics of the Japanese transistor market. One is that transistors were used for commercial goods, not military applications as was the case in the U.S. And the other is that the commercial application encouraged the entry of many companies into the transistor market and stimulated market competition. One of the executives of Company B expressed his view that commercially used semiconductors actually require much tougher and higher quality standards than those in military use. Military semiconductors are used in a well-controlled environment, free from dust and at a constant temperature, while commercial semiconductors have to be operable under many different daily living conditions. Moreover, Japanese consumers are extremely attentive not only to quality but also to price. These difficult market conditions have led to severe product-market competition in Japan.

SONY's innovation did not mean that Japanese companies were significantly competing against Western producers in this early stage of the semiconductor industry. But given the technological capability and national economy, the market was highly competitive for them. The government's infant industry approach restricted foreign investment, disseminated technology to a wider body of companies, and stimulated indigenous technology development in the electronics industry. This approach not only gave some additional time to Japanese companies for developing their technology, but also stimulated competition over foreign licensing among companies that had very similar levels of technological capability.

Although MITI had been known for its tight control over technological licenses until 1968 (Appendix Table 2.3), the government was not quite successful in maintaining its original policy stance in the case of the semiconductor industry. Market competition was already severe among six starting companies (SONY, Hitachi, NEC, Toshiba, Kobekogyo, and Fuji Electric), though product markets were expanding a little. In 1959 four companies (Mitsubishi Electric, Japan Radio, Oki Electric, and Sanyo) ignored MITI's warning, entered the transistor business, and further aggravated competition in the market. The commercial applications of transistors became so extensive that even specialized companies expanded the scope of transistor production to other commercial goods. For example, Company B, which had been specializing in computer and communication equipment, changed its corporate strategy in 1958 to produce transistors for use in non-specialized commercial goods, thus enabling the company to take advantage of scale merit. Consequently, severe product-market competition drove companies to differentiate their products by quality, reliability, stability, and price and to search for more product markets in foreign countries (Ohno and Kawakatsu, 1983).

Product differentiation was not yet achievable through R&D, due to the immaturity of technological capability. Instead, raising the quality of skilled and unskilled workers was perceived to be the primary target for improving product quality as well as cutting production costs, especially since production was quite labor-intensive. For example, Company B established a training center within its factory as early as 1956, and then it opened a technical school in 1960. Company A conducted the first research report presentation in 1959 and also established a training school in 1960.

Consequently, the production of the germanium transistor increased from close to zero yen in 1956 to about 15 billion yen in 1959 (Denpa Shinbunsha, 1969:17). By the end of this period, it was said that the quality of Japanese rate grown transistors outstripped that of the U.S. products (Nakagawa, 1985). Japan became the world's largest transistor producer in 1959, and even the U.S. started restricting imports from Japan. Nevertheless, this does not mean that Japan caught up with the U.S. technologically. Development of the integrated circuit by American companies forced the Japanese to start all over again from scratch.

Thus, in the imitation-for-production period, Japanese companies came to mass-produce transistors based on foreign technology with little R&D of their own. However, from the very beginning they faced severe domestic market competition, which compelled them to improve the reliability, stability, and quality of their products and to export as many as possible. Their survival strategies resulted in the country's status as the world's largest transistor producer in 1959, but Japanese companies accomplished this without the basic innovative capability.

3.   Innovative Imitation Period (1960 - 1964)

The innovative imitation period began with IC research by NEC and Mitsubishi in 1960 and a consequent boom in newly-established research institutes among electronics-related companies in 1961. The major purpose of increasing research was to catch up with the U.S. in IC and silicon transistor technologies. Although their research on IC remained in the experimental imitation stage, studies on silicon transistors reached a much more advanced stage. Even so, silicon transistor production faced many serious problems. Silicon and zinc required, respectively, very high and ultra high levels of purity, and a stable environment in a factory with very strict temperature, moisture, and dust control was necessary. These problems tormented researchers for quite a long period of time. As a matter of fact, the Japanese development of the silicon planer transistor in 1962 lagged three years behind the U.S. innovation. Japanese success was again greatly owing to a foreign technology license, which enabled SONY to develop the silicon epitaxial planar transistor for high voltage television (Nakagawa, 1985).

The biggest highlight of this period was the world's first transistorized calculator produced by SONY and Sharp. Another important contribution was Hitachi's development of low temperature passivation technology and two other processing technologies in 1963, which later relieved Japanese companies from their dependence on Fairchild's planar technology (Appendix Table 2.3).[4] The appearance of some significant innovations in products and processing technologies signified that transistor technology was setting its roots in the Japanese industry.

The spread of semiconductor technology allowed even smaller and specialized companies to venture into transistor and diode production with far less risk than had been necessary in an earlier period. Though transistor and diode market expansion was gradual between 1960 and 1964, seven companies (Shindengen Electric, Sanken Electric, Origin Electric, Hachio Electric, Sansha Electric, Meidensha, and Japan Storage Battery) contracted technical agreements with U.S. companies and entered the market. This made market competition very severe.

The increased technological capability of Japanese companies and intensified product-market competition created a new direction in corporate strategies, namely, increasing efforts for automation and rationalization. For example, Company A placed a very strong emphasis on the rationalization of production processes in 1960, on technological self-reliance in 1962, and on catching up with the U.S. technological capability in 1963. Furthermore, facing a sluggish economy in 1964, the company emphasized increasing competitiveness by improving product quality and reducing production costs.

---

[4] Since NEC was the sole licensee of Fairchild's planar technology in Japan, Hitachi had to obtain the license for one of the most essential technologies of semiconductor production from its major competitor in Japan. Hitachi's success in developing its own low temperature passivation technology released the company from its dangerous dependence on its major competitor (Aida, 1992:48).

These corporate goals resulted in strategies such as quality circle promotion, soliciting suggestions for cost reduction, automation of production processes, and production cost cutting activities. Replacing foreign technology licenses with indigenous technology was one of the most important means of cutting cost. Promoting intensive cooperation and increasing quality and cost control with suppliers was another important strategy.[5] From the middle of the 1960s, developing CCC interfirm relations became a very important part of corporate strategies. This era marks the beginning of developing CCC interaction in the Japanese semiconductor industry.

Thus, with regard to transistors, this period was a very innovative period, characterized by such market-driven innovations as transistor radios, micro-televisions, and transistorized calculators. Although Japanese companies were still heavily dependent on foreign patents, a few indigenous foreign-patent-replacing innovations started appearing. These technological developments were partly owing to severe product-market competition, which was intensified further by the entry of new and specialized companies. To reduce patent and technical-fee payments meant a significant reduction of production costs.

C. Development of the Integrated Circuit (IC) and Other Semiconductors

1. Experimental Imitation Period (1958 - 1962)

The next generation of semiconductor technology started appearing when Westinghouse Electric and RCA announced two innovations in 1958: Westinghouse had developed the molectronics and RCA produced the micromodule. And the announcements of the solid state circuit by Texas Instruments in 1959 and the monolithic IC by Fairchild in 1961 produced even bigger shocks to Japanese companies. The basic idea behind these developments was to eliminate the connecting points between transistors that had become a frequent source of troubles and failures. The U.S. Department of Defense initiated the IC research to develop a light and reliable computer for use on space projects, and they provided research labs with a large amount of money.

Despite Japanese success in transistor production and some innovative products and processing technologies, these U.S. innovations threw Japanese companies back to the experimental stage. In attempts to catch up with

---

[5] Interviews with managers of companies A and B.

this advancement, NEC and Mitsubishi announced the beginning of their own research on the IC in 1960, and many semiconductor-related companies started establishing research institutes in 1961. In Company B, a top executive emphasized the importance of IC research and developing communication equipment and computers with ICs. However, despite the enthusiasm and expressed corporate goals, researchers faced many difficulties due to the lack of basic technological capability.

As TRI had at the very beginning of transistor research, the Electro-Technical Laboratory (ETL), another government research institute under MITI, successfully produced the first experimental model, building a solid state circuit in 1960 without foreign assistance (Nakagawa, 1985). It is interesting that these experimental periods were initiated by the government research institutes, TRI and ETL, without foreign assistance.

However, when it came to the stage of mass production, Japanese companies had to rely on foreign technology licenses. Until NEC contracted an agreement with Fairchild on the planer patent in 1962, all Japanese companies were struggling without the silicon oxidized film, epitaxial, or planer technologies. Fairchild's planar patent enabled NEC and Mitsubishi to finally succeed in the sample production of the IC in 1962.[6] Thus, U.S. development of the IC created a fear in Japan that Japanese technological accumulation was not sufficient enough to keep up with the world technological level, and that Japan had to start all over again from licensing agreements.

2.   Market Uncertainty Period (1963 - 1964)

Germanium transistor sales became more important than the sales of vacuum tubes in 1961, and they kept up an ascending path until 1964. In 1960, silicon transistors appeared in the product market, and kept increasing their market share even after the share of germanium transistors started declining in 1965. Given the booming sales of germanium transistors and the increasing sales of silicon transistors between 1963 and 1964, companies found it difficulty to sell ICs (Denpa Shinbunsha, 1969:17).

Because of the strong position of U.S. ICs, Japanese engineers predicted that the major product in the coming markets for Japanese producers would be the custom-designed silicon transistor. Even so, engineers faced resis-

---

[6] Without personal relations between Mr. Osafune of NEC and Mr. Noyce of Fairchild, the time lag of IC production could have been more than one year. In fact, the development of an earlier version of an IC, the molectronics, lagged about two years behind (Appendix Table 2.2).

tance against new IC research projects from management due to the heavy investments already made in semi-automated germanium production facilities (Ohno and Kawakatsu, 1983). This meant that Japanese companies' efforts to improve germanium transistor production turned out to retard the advance into silicon transistors and ICs. Besides, they did not have enough confidence to make investments in IC technology and compete against U.S. IC imports.

Japanese ICs were not attractive in product markets partly because the main domestic IC users, computer and telecommunication equipment manufacturers, constituted only a small segment of the electronics equipment and home appliance markets, and partly because these markets were dominated by less expensive imported U.S. ICs (Ohno and Kawakatsu, 1983). Understandably then, Japanese companies were quite hesitant to advance to the IC. Meanwhile, foreign producers made significant technological progress with such inventions as the metal-oxide-semiconductor integrated circuit (MOS IC) by Texas Instruments and General Instrument, solid logic technology for complete production automation by IBM, and silicon-on-sapphire technology.

3.   Innovative Imitation Period (1965 - 1967)

The movement of the IC market was finally stimulated by Texas Instruments' announcement of investment in Japan in 1964 (Appendix Table 2.3). Since the possibility of American monopoly in the market, a domination based on superior technology, was quite alarming to Japanese companies, they started shifting strategies from the transistor to the IC in 1965. Never before in the history of the Japanese semiconductor industry, had the entry of a single company into the market threatened Japanese companies and stimulated their R&D activities.

The primary target of Japanese research was the development of the MOS IC, which was free from many of the problems that plagued the molectronics, the micromodule, and the solid state circuit. Besides, the MOS IC had a strong potential to expand the density of information contained in a single unit of memory (Nakagawa, 1985). In 1965, Company A heavily emphasized the development of IC and IC-using products in order to prepare for the coming competition against Texas Instruments, and against domestic competitors. Company B also made drastic management reforms, moving toward the internationalization of management and the reorganization of the R&D system. The company implemented management improvement activities in 1965, and it established the Quality Control Department and the Reliability Control Department directly under the President in 1967, in order to continuously improve the quality and reliability of products.

In 1965, Texas Instruments was forced by the Japanese government to open its patents to the public in order to obtain permission to invest. The opened patents provided Japanese companies easier access to superior foreign technologies related to PN isolation, diffusion, and ion implantation (Nakagawa, 1985), and the legal action simultaneously prevented Texas Instruments from exercising unfair control by refusing Japanese use of its patents (Denpa Shinbunsha, 1968). These patents, along with Hitachi's MOS FET technology in 1963, significantly helped Japanese companies to produce the first MOS IC in 1967, two years behind the U.S.

One significant development of this period was the world's first all-IC calculator, a calculator developed in 1967 with more sophisticated capabilities than U.S. machines. Moreover, research begun in the early 1960s had started blooming, with patents on process technologies such as metal oxide passivation, multiple masking, MOS transistor protection, $SiSiO_2$ metallic oxide filming, $SiSiO_2$ aluminophosphide glass surface processing, the-world's first electron beam lithography, and ion implantation. Furthermore, low temperature passivation technology was improved in a way that overcame the weaknesses of the planar technology and released Japanese companies from their dependence on the Fairchild patent. Hence, this period can properly be called the stage of innovative imitation. This period was also one of application engineering, for companies applied existing technologies to new products with an unprecedented degree of development and accumulation of indigenous technology. This quick advancement could not have been achieved without Japanese companies' cumulative efforts to develop indigenous technology or without Texas Instruments' threat to their survival in the Japanese market.

4.   Application Engineering Period (1968)

The application engineering period was the period that introduced Japanese leadership in the world memory IC market. The N-channel MOS IC, originally developed in the U.S., had so many problems that U.S. companies could not commercialize it, despite its speed which was faster than the P-channel MOS IC. Young researchers in Japanese companies, however, predicted that it would be the IC of the future (Nakagawa, 1985), and persuaded the government to select it as the core semiconductor for development in the government's high capability computer project (the Chokoseino Denshi Keisanki Purojekuto). Three companies cooperated under the project and successfully developed the world's first N-channel 144-bit MOS memory IC.

This development gave Japanese companies some technological advantages over U.S. producers in the area of memory IC. For the first time, R&D became more technology-driven than market-driven. This advancement into the application engineering stage was made possible by the unprecedented development of indigenous processing technology, which gave domestic technology a more dominant role than foreign ones. This development was also indispensable to companies that wanted to cut into the U.S.-monopolized Japanese IC market and to seriously compete against the technologically advanced Texas Instruments.

Technological success and a favorable balance of trade gave the Japanese government the confidence to allow a joint venture between Texas Instruments and SONY and to lift control over technical agreements for the semiconductor industry.

However, inexpensive U.S. ICs continued to flood the Japanese market. Between 1965 and 1969, virtually all ICs consumed in Japan were imported, and almost all came from the U.S. (Denpa Shinbunsha, 1977:643). What was worse, in 1968 Texas Instruments succeeded in producing the world's first MOS LSI, advancing one step beyond Japanese technology level. Furthermore, IBM began research on the Josephson junction, which was considered the base technology of the next generation of ICs. Thus, U.S. market domination and technological advancement prevented Japanese companies' success with the N-channel MOS IC and denied them the chance to strengthen their market position.

D. Development of the Large-scale Integrated Circuit (LSI) and Other Semiconductors

1. Application Engineering Period with Market Uncertainty (1969 - 1971)

Texas Instruments' development of the LSI threw Japanese companies back to the starting line. When Japanese semiconductor companies were working on the transistor and IC, technical agreements were the way to begin developing technological capability. This time Japanese campanies tried to develop their capability without foreign licenses. The process of developing LSI technology, however, was quite difficult. The LSI required highly sophisticated manufacturing equipment with system designing capability. The Japanese companies' lack of experience in both of these fields put them through a period of great difficulty and meant that intensive R&D was required. Japanese researchers found that a presumably clean room was not high grade enough. And the passivation of LSI surfaces was not tight enough to keep dust and water away. Even so, two years after the first U.S. LSI, Japanese companies succeeded in producing the MOS LSI and even introduced the world's first 2048-bit MOS LSI in 1970. Toshiba was also successful in developing the perfect crystal and chemical vapor deposition technologies, which allowed the company to be free from the planar patent. Major semiconductor producers began to develop their own process technologies in order to be self-reliant and cut patent costs. Unlike the periods of transistor and IC development, the period of LSI development involved extensive use of indigenous technology. Hence, this period was an application engineering period, because domestic technologies enabled companies to engage in innovation more technology-driven than market-driven.

But this success did not guarantee Japanese companies an increased share in the LSI market. To the contrary, the domestic market was more flooded with inexpensive U.S. imports than ever before, causing market uncertainty for Japanese companies. A leading Japanese calculator producer even started importing MOS LSIs from the U.S.

To make the situation worse, due to rapid expansion of the IC market, former customers of major Japanese semiconductor manufacturers started contracting technology agreements with U.S. companies and producing by themselves (Asahi Glass, Nippon Columbia, Mitsumi Electric, Yamaha, and Toko in 1969 and Canon in 1971).[7] Even though some of them obtained only IC technology for internal parts supply and did not pick up LSI technology, the termination of business with former customers as well as their potential for future participation in the market undeniably stimulated competition.

Given product market uncertainty and intensified competition, Company B strengthened QC activities to achieve zero-defect levels of their products, and it started teaching young engineers the basis of system and circuit design (Nakagawa, 1985). Similarly, in 1969, Company A emphasized increased international competitiveness in IC production and strengthened IC and LSI development capability by establishing a new department in 1972. For improving worker capability, it also established a lecture series on IC technology for skilled workers in 1969 and a training factory in 1970.

However, around 1970 a turnaround took place. Calculators with imported U.S. ICs and LSIs malfunctioned so frequently that they severely damaged the reliability image of U.S. products (Nakagawa, 1985). Even foreign buyers came to recognize the higher reliability and quality of Japanese prod-

---

[7] Kokusai Electric and Toyo Electric Seizo also contracted technical agreements for producing semiconductor manufacturing equipment.

ucts and purchased them. Consequently, some of the major U.S. electronic companies started withdrawing from the memory IC and LSI markets in 1971 and Japan experienced a drastic reduction in import dependence by 1972 (Denpa Shinbunsha, 1977:643).

This turnaround is indicative of the strategic differences between U.S. and Japanese companies. Both started feeling the squeeze of the market in the early 1970s due to higher labor costs and stronger customer demand for a less expensive semiconductor. In addition, the Japanese companies faced more severe product-market competition and uncertainty due to the entry of foreign and new domestic manufacturers. U.S. companies addressed their problems by investing in developing countries in order to make use of an inexpensive labor force. According to a Japanese interviewee, this was the major cause of U.S. failure. Japanese companies hesitated to move into developing countries, since control over personnel quality and the training of workers would be difficult in a distant place, and since CCC interaction could not be fully utilized to bring improvements in technology, quality, costs, delivery, and services. For them, foreign investment meant the production of unreliable, high cost, and low quality products.

Instead, they implemented diverse strategies to cope with their problems. Internally, they automated assembly processes, promoted zero-defect movements, familiarized both skilled and unskilled workers with the highly sophisticated processes of semiconductor production, and solicited worker participation in improving the production process. Externally, they enhanced CCC interaction with equipment manufacturers and parts and material suppliers. They mutually exchanged information, searched for ways to reduce production costs, ensured the high quality and reliability of final outputs, coordinated delivery and other activities, and jointly developed technology.

One important Japanese strategy was to strengthen CCC interaction with cooperating companies as a long-range means of improving technology, quality, cost, delivery, and services. Decision makers hoped that CCC interaction would also help them to create innovative technology and to survive in highly-competitive product markets (Ohno and Kawakatsu, 1983). Hence, more CCC-type relations came into existence as indigenous technology became more dominant. These corporate strategies significantly helped Japanese companies to increase their comparative advantages vis-á-vis U.S. companies.

2.   Application Engineering Period (1972 - 1974)

A far more turbulent product-market environment began around 1972. Several factors caused the turbulence. First, five new Japanese companies (TDK,

Kishimoto Sangyo, Alps, Hitachi Chemical, and Osaka Titanium) entered the semiconductor business.[8] Second, Texas Instruments withdrew from a joint venture with SONY in 1971 and started operating independently. Third in 1971, the Japanese government started accepting foreign ownership of up to 50 percent of companies producing ICs with less than 50 elements. Fairchild and Motorola established joint ventures with TDK and Alps respectively in 1972 and 1973, which meant that the world's three biggest semiconductor companies started producing in Japan. After the government lifted control over the importation of IC units and liberalized foreign ownership of companies producing ICs (everything but computer use was allowed in 1974), Motorola and Fairchild dissolved joint ventures and went into independent operations.

Such aggravated competition stimulated Japanese companies to continue their efforts, begun in the 1960s, to strengthen their competitive edges. They again made strong efforts to automate production and strengthen quality control. For example, Company A established a management improvement program in 1973, called "MI Challenge 50." This program was designed to increase the number of patents, reduce unproductive projects, enable quicker technology development and customer claim responses, simplify and shorten production processes, and reduce the amount of defective goods, paperwork, time spent in meetings, and accidents. Company B extensively expanded computer-automated administration and IC and LSI production in 1973. It also established such activities as "Spark 10 Activities" (1971) to reduce production costs by ten percent and "Quality Strategy Activities" (1974) to improve the quality of management, product, service, environment, public relations, and company image (See Appendix Table 2.5).

Japanese companies intensified their R&D, and developed processing technologies such as LSI mask defect detection, completely automated assembling, mass-production-purpose tape careers, and ion implantation. NEC was also successful in developing the 1K-bit N-channel MOS LSI (1972) with hardly any lag time after IBM, indicating that Japan had reached an essentially equal technological level in the area of memory IC. In developing the 4-bit microprocessor, Japanese companies were two years behind, but they were not behind at all in the 16-bit microprocessor. Japanese companies had essentially caught up technologically in the memory IC area. The achievement was very important, since memory IC had more value added than any other device, and since this area involved the most advanced tech-

---

[8] Daini Seikosha also contracted a technical agreement on germanium manufacturing equipment in 1972.

nology that could be used for producing other devices. Thus, Japanese companies became leading figures in a highly profitable area, despite severe market competition, by emphasizing continuous production improvements, strengthened technological capability, and intensified R&D.

E.   Development of the Very-large-scale Integrated Circuit (VLSI) and Other Semiconductors - Leveraged Creativity Period (1975 - 1983)

During the leveraged creativity period, Japanese companies established their technological leadership in the area of memory IC. U.S. and Japanese companies produced 16K-bit MOS LSI in the same year (1975). But, owing to joint research organized under the VLSI Cooperative, Japanese companies succeeded in advancing processing technology significantly and producing the world's first 64K-bit MOS VLSI in 1977.

Since then, Japanese companies have been the world's first in the production of 128K-bit (1978) and 256K-bit MOS VLSIs (1980). In the 256K-bit MOS VLSI, they were ahead of U.S. companies by roughly two years. For the first time, Japanese IC trade recorded a world trade surplus in 1979, and even had a surplus in trade with the U.S. in 1980. Japanese companies held 70 percent of the 64K-bit DRAM world market in 1979, 44 percent of the 16K-bit DRAM world market in 1982, and 90 percent of the 256K-bit DRAM world market in 1984.

In addition to taking the leadership of the memory IC market, Japanese companies began paying serious attention to forefront technologies in other types of semiconductors. They started research on the Josephson junction in 1975, seven years behind the U.S. Ironically, when IBM withdrew from this area of research in 1983, the Telecommunication Research Institute of the Ministry of Post and Telecommunication (TRI) succeeded in developing the Josephson 1K-bit memory IC (Appendix Table 2.2). This strategic difference allowed Japanese companies to advance ahead of the U.S. in Josephson technology.

The world's first developments in frontier technologies also began in this period: Fujitsu succeeded in inventing the high electron mobility transistor (HEMT) in 1980; TRI produced the gallium arsenide (GaAs) 1K-bit SRAM IC in 1982; TRI and Nikon cooperatively manufactured the x-ray lithography system in 1978; and Sumitomo Electric Industries built the epitaxial wafer mass-production technology for GaAs IC in 1979. However, U.S. companies were still ahead of Japanese companies by about a year with products including microprocessors and single chip microcomputers. But even though the U.S. maintained leadership in some fields, many of the Japanese innovations were radical enough to demonstrate that network- and cooperation-oriented Japanese companies had the capability to make radical innovations.

Table 2.1 compares the technological level of U.S. and Japanese semiconductor industries at the beginning of the 1980s. The U.S. was advanced in all design technology areas (system design, CAD technology, and CAD equipment) and also in terms of device structure (new isolation technology and devices such as the silicon gate array, LOCOS, and FAMOS). These were areas where individual creativity was of the utmost importance for development. In contrast, Japanese companies were strong in production areas where organizational capability could be fully used. They were more advanced in sub-micron processing technology and process automation technology, and they were at an equal level in oxidation, diffusion, chemical vapor deposition, and other vapor deposition technologies.

The strength of Japanese companies in production processes is shown in Table 2.2, which compares the quality of 16K-bit DRAM LSI among six Japanese and U.S. companies. As the table shows, even the best U.S. company could not reach the quality level of the worst Japanese company, and the worst U.S. company recorded a score close to a half of the best Japanese company. These figures clearly indicate the superiority of Japanese product quality as well as production process technology at the beginning of the 1980s. Their success resulted partly from R&D, quality improvement activities, high annual average of plant-and-equipment-investment growth rates, and the development of CCC interaction.

Table 2.3 shows that Japanese companies started providing technologies to foreign companies, especially information on the production of memory IC, while they still received technologies related to the microprocessor, gate array, CMOS, CAD, CMOS, EPROM, and EEPROM. What Japanese companies lacked was individual creativity and system designing capability, and they imported U.S. technology related to products that involved complex system designing. Company A's corporate decision in 1976 to develop an indigenous microprocessor and cultivate capability in system designing represented executives' awareness of Japanese companies' problems.

The development of international technological interdependence was reflected in the balance of payments for patent and technical fees (Figures 2.2 and 2.3). In the 1980s, Japanese companies started to reach a balance between payments and receipts for patent and technical fees. Company A balanced around 1980, while Company B – although statistics are available only for the whole company – balanced in 1983. Since then, receipts have

Table 2.1 Comparison of Technological Levels Between U.S. and Japanese IC Industries at the Beginning of the 1980s

TECHNOLOGY	JAPAN	COMPARISON	U.S.	COMMENTS
A. Design Technology				
1. System Design	Still weak in areas that require creativity, but strong in areas where goals are well established, owing to strong organizational capability	<	Strong in areas where individual creativity is required	U.S. is advanced in the microprocessor, but Japan is advanced in the memory IC
2. CAD Technology	Many researchers engage in LSI design, creating a shortage of researchers in developing CAD technology	<	Advancing rapidly in both CAD hardware and software	U.S. is advanced, but Japan is catching up in some areas
	At the forefront in the pattern testing CAD system		At the forefront in software such as SPICE for simulation, and STICKS and CABBAGE for symbolic design	Japanese companies study and improve U.S. technology for practical use
	Beginning of R&D in simulation and symbolic design CAD systems			
3. CAD Equipment	Beginning of domestic production	< <	At the forefront technology in the production of CAD equipment (CALMA, APPLICON, etc.)	U.S. far advanced
B. Production Technology				
1. Sub-micron	Mass production at 2 micron level and R&D at 1.2 to 1.5 micron level	≥	Mass production at 2.5 to 3.0 micron level and R&D at 1.2 to 1.5 micron level	
	Advanced in electron beam mask lithography and dry processes (RIE, plasma, etc.)		Advanced in basic research on X-ray and ion beam lithography, but only a few companies use advanced technology	
2. Oxidation, Diffusion, and Chemical Vapor Deposition (CVD) Technologies	Advanced in low-defect thin film deposition technology (low-defect CVD thin-film, aluminum thin-film, etc.)	=	Advanced in thin film deposition technology with new materials (P-Sio thin film, $MOSi_2$ thin film, etc.)	Japan is advanced in low-defect technology, but U.S. is advanced in new material technology
3. Overall Process a. Device Structure Technology	Advanced in new ideas on isolation technology Some technologies are already in use	<	New isolation technology for VLSI reported by IBM, BTL, HP, etc.	U.S. has been leading (Si gate array, LOCOS, isoplanar, FAMOS, etc.)
b. Process Automation	Advanced in actual production management, though behind in production process design	≥	Many companies are far behind, but IBM is advanced	
4. Packaging and Bonding Technology	Advanced in packaging and bonding technologies for mass production	=	Advanced in packaging technology to produce ICs with ultra-high reliability (space and military use)	

Note: > > Far more advanced
    > More advanced
    ≥ Slightly more advanced
    = Equal
    < Less advanced

Source: Denpa Shinbunsha, *Denshi kogyo nenkan 1988* (Yearbook of electronics industry), (Tokyo: Denpa Shinbunsha, 1988), 780.

been higher than payments. Figure 2.2 shows the l960s as a period of "importation," the 1970s as a time of "cross-licensing" and the 1980s as a period of "exportation." These figures indicate that Japan was successful in developing a technological system of its own and achieved a state of technological self-reliance in the beginning of the 1980s.

Given the dynamics of the industry, eight additional companies (Nippon Denso, Clarion, Suwa Seikosha, Ricoh, Olympus, Asahi Chemical, Yamaha,

Table 2.2 U.S.-Japan Comparison of 16K-bit DRAM Quality

COMPANY	DEFECT RATE OF SAMPLE	DEFECT RATE IN THE FIELD (PER 1000 HRS)	OVERALL QUALITY INDEX
J1	0.0%	0.010%	89.9
J2	0.0	0.019	87.2
J3	0.0	0.012	87.2
U1	0.11	0.059	86.1
U2	0.19	0.090	63.3
U3	0.19	0.267	48.1

Notes: J: Japanese Firm; U: U.S. Firm
Source: Reported by Hewlett-Packard Company quoted in Denpa Shinbunsha, *Denshi kogyo nenkan 1981* (Yearbook of electronics industry), (Tokyo: Denpa Shinbunsha, 1981), 702.

Table 2.3 Technical Cooperation in Semiconductor Industry

	DIRECTION OF TECHNOLOGY FLOW	
	JAPAN	U.S.
	Hitachi ⟶	Hewlett Packard (64K-bit DRAM technology)
	Toshiba ⟶	Zilog (64K-bit DRAM sub-micron processing technology)
	Toshiba ⟵	Zilog (microprocessor & microcomputer technology)
	Toshiba ⟷	LSI Logic (gate array, CMOS process and CAD technology)
	Oki ⟶	National Semiconductor (64K-bit DRAM technology and joint developments)
	NEC ⟷	AMI, Zilog (microprocessor technology)
	Oki, Fujitsu ⟵	Intel (microprocessor technology)
	Fujitsu ⟶	MMI, TI (gate array technology)
	NMB ⟵	INMOS (256K-bit DRAM technology)
	SONY ⟵	Vitelic (256K-bit DRAM technology)
	Sharp ⟵	Wafer Scale Integration (CMOS EPROM technology)
	Oki ⟵	Exell (EEPROM technology)

Note: Oki: Oki Electric Industry Co. Ltd.
           TI: Texas Instruments, Inc.
           MMI: Monolithic Memories, Inc.
           Exell: Exell Microelectronics Inc.
Source: Denpa Shinbunsha, *Denshi kogyo nenkan l986* (Yearbook of electronics industry), (Tokyo: Denpa Shinbunsha, 1986), 7 14.

and Fuji Xerox) entered the semiconductor market. They were all former customers of semiconductor manufacturers. These dynamics also led to U.S.-Japan trade disputes over the semiconductor, compelling Japanese companies to invest in foreign countries. There is abundant evidence from this period to show that the technological level of the Japanese semiconductor industry was already at the leveraged creativity stage, even though Japan was still a long way from the state-of-the-art breakthrough stage.

Figure 2.2 Balance of Payments for Patent and Technical Fees in the Semiconductor Division of Company A (1966-1984)

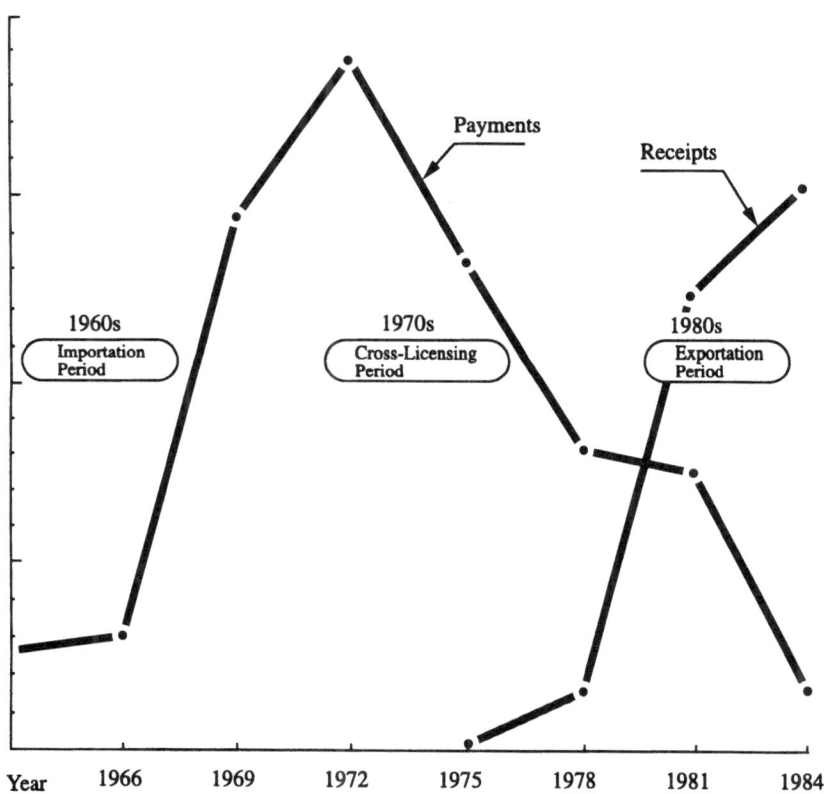

Source: Company A

Figure 2.3 Balance of Payments for Patent and Technical Fees in Company B (1980-1985)

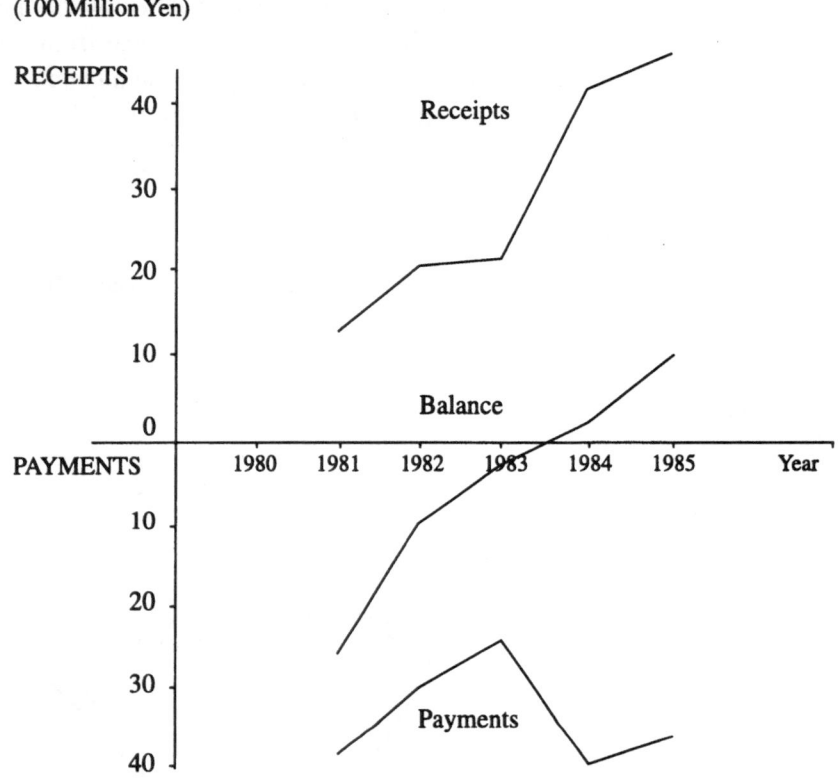

Source: Company B

F.   Development of Ultra-large-scale Integrated Circuit (ULSI) and Other Semiconductors - Leveraged Creativity Period (1984 - 1992)

During the leveraged creativity period, Japan established a remarkable position in the world. It increased its share of the world semiconductor market from 27% in 1980 to 49.5% in 1990, while the U.S. and Europe declined from 57% to 36.5% and from 16% to 10.5% respectively (Press Journal, 1987:62; Nikkan Kogyo Shinbun, 1991c). Japan's world IC market share increased from 25.1% in 1978 to 51.0% in 1988, while the U.S. declined from 65.9% to 36.5%. Europe increased its share by a small margin (Denpa

Shinbunsha, 1981:690, 1990:763).

According to Table 2.4 which shows the ranking of the top ten semiconductor manufacturers, the top two positions were maintained by Texas Instruments and Motorola until the mid-1980s, but since 1986, NEC, Toshiba, and Hitachi have occupied the top three positions. These three Japanese companies had 23.6% of the world semiconductor market in 1991, while the top three U.S. companies, Motorola, Texas Instruments, and Intel, had 16.1%.

Along with its increased share in the world semiconductor market, Japanese companies developed technologies in diverse devices, recording many firsts. They broke ground in the memory IC field by developing 4M-bit DRAM in 1986, 16M-bit DRAM in 1987, and 64M-bit DRAM in 1990. Another area of strength was their development of devices with new materials. They produced the world's first HEMT 1K-bit SRAM in 1984, and continued with firsts including HEMT 4K-bit SRAM in 1987, HEMT 64K-bit SRAM in 1990, a 1-chip bio sensor in 1984, a Josephson chip for multiplication in 1984, 1K-bit lead-based Josephson memory in 1985, a single

Table 2.4 Semiconductor Manufacturers Ranked by World Market Share in 1971 and from 1980 to 1991

COMPANY	'91	'90	'89	'88	'87	'86	'85	'84	'83	'82	'81	'80	'71
		(1990 WORLD MARKET SHARE)											
NEC	1	1 (8.5%)	1	1	1	1	1	1	3	3	3	4	6
Toshiba	2	2 (8.4)	2	2	2	3	5	5	5	5	5	6	
Hitachi	3	3 (6.7)	3	3	3	2	4	4	4	4	4	5	7
Motorola	5	4 (6.3)	4	4	4	4	2	2	2	2	2	2	2
INTEL	4	5 (5.4)	8	7	8	11	8	8	7	7	7		
Fujitsu	6	6 (5.2)	5	6	6	6	7	9	8	8		7	
TI	7	7 (4.4)	6	5	5	5	3	3	1	1	1	1	1
Mitsubishi	8	8 (4.2)	7	8	9	10	13	13	13	11	9		
Matsushita Electronics	9	9 (3.3)	9	9	10	9	10	12	9	9	8	8	
Philips*	10	10 (3.3)	10	10	7	8	6	6	11	11	9	3	
National Semiconductor	11						9	7	7	6	6	10	4
Information Source**	!	@	@	@	#	#	%	%	&	&	$	&	$

Note: * Signetics (U.S.) is included in Philips.
    ** All information originally was compiled from Dataquest reports.
Sources: @ Press Journal, *Nihon handotai nenkan 1991* (Japan semiconductor yearbook), (Tokyo:Press Journal, 1991), 49.
    % _____, *Nihon handotai nenkan 1987* (Japan semiconductor yearbook), (Tokyo: Press Journal, 1987), 199.
    & _____, *Nihon handotai nenkan 1985* (Japan semiconductor yearbook), (Tokyo: Press Journal, 1985), 104, 512.
    # Denpa Shinbunsha, *Denshi kogyo nenkan 1989* (Yearbook of electronics industry), (Tokyo: Denpa Shinbunsha, 1989), 764.
    $Nihon Denshi Kikai Kogyo Kai (Electronics Industries Association of Japan), '91 IC guidebook, (Tokyo: Nihon Denshi Kikai Kogyo Kai, 1991), 17.
    ! Nikkei Sangyo Shinbun, "*91nen handotai seisangaku, INTEL Bei shui ni, sekai rankingu 4 i ni fujo* (91 Semiconductor production, INTEL U.S. No. 1, world No. 4)." January 8, 1991, 6.

phase yttrium barium copper oxide Josephson device in 1987, a Josephson junction microprocessor in 1988, a Josephson computer in 1989, Josephson 8-bit digital signal processor in 1990, a GaAs logic LSI in 1985, a GaAs hetero bipolar transistor with complete monolithic IC in 1988, and 1K-bit optical memory with a vertical-to-surface transmission electron photonic device in 1988 (see Appendix Table 2.1). Many of these technologies can be considered radical innovations. U.S. companies, however, continuously showed technological leadership with their silicon processors and micro-computers. For example, they developed the world's first 32-bit micropro-cessor in 1982, reduced instruction set computer (RISC) processor in 1986, and 32-bit GaAs microprocessor and 64-bit microprocessor in 1989 (Ap-pendix Table 2.2). Although Japanese companies have been quite active in R&D in new devices with new materials, none of these innovations can be classified as breakthroughs.

At the beginning of the 1980s, U.S. technology showed overall superior-ity in the semiconductor industry. But in the latter 1980s, Japanese compa-nies became stronger and were still improving their positions. According to Table 2.6, Japanese companies were leading in DRAM, SRAM, and bipolar silicon devices, while U.S. companies were leading in custom products and

Table 2.5 Top Ten Semiconductor Equipment Manufacturers by Sales in Selected Years

COMPANY	'89 RANK	'89 SALES ($ MILLION)	'88	'83	'82	'79
Tokyo Electron (J)	1	633.9	2			
Nikon (J)	2	587.2	1	8	10	
Applied Materials (U)	3	523.3	4	7	5	3
Advantest (J)	4	398.8	3	9	4	
Canon (J)	5	383.6	6	10	8	
General Signal (U)	6	353.7	5	6	9	
Varian (U)	7	335.0	7	4	2	6
Hitachi (J)	8	210.0	11			
Teradyne (U)	9	199.9	9	5	7	5
ASM International (U)	10	186.8	13			
Perkin Elmer (U)				1	1	2
Fairchild Test (U)						
Systems (U)				2	3	1
GCA (U)				3	11	4
Eaton (U)				6	6	8
Schlumberger (U)					3	
TEKTRONIX (U)						7
Kulicke & Soffa Industries (U)						9
Information Sourc	$	$	$	*	%	@

Note: (J): Japan, (U): U.S.
Sources: $ Nihon Denshi Kikai Kogyo Kai (Electronics Industries Association of Japan), '91 IC guidebook, (Tokyo: Nihon Denshi Kikai Kogyo Kai, 1991), 29.
         @ Press Journal, *Nihon handotai nenkan* (Japan semiconductor yearbook), (Tokyo: Press Journal, 1990), 77.
         * Takeo Shimura,T, *IC sangyo no shintenkai* (New developments of IC industry), (Tokyo: Diamond, 1984), 216.
         % Nihon Keizai Shinbun, "*Nichibei sangyo kyozon e no michi* (Way for cooperation between Japanese and U.S. industries)," May 8, 1990, 10.

microprocessors. With non-silicon devices Japan was leading in all devices except linear. With regard to materials, Japanese companies were leading in both silicon and gallium arsenide. In production technologies, they were strong in furnace, packaging, and testing system technologies, while U.S. companies led in electron beam lithography and stood equal in vapor deposition, diffusion, assembling, and CAD technologies. The U.S. was strong in areas that involved highly advanced and crucial technologies. Although the countries were in an equal position in optical and x-ray beam lithography and CAM technologies, Japanese companies were becoming stronger in these technologies. Overall, the Japanese had established superiority in semiconductor technology, though U.S. companies maintained their strength in technologically sophisticated areas.

Japanese semiconductor materials manufacturers also became quite dominant in the world market in the 1980s. Moreover, there is abundant evidence that Japanese semiconductor materials companies increased their market

Table 2.6 Relative Japanese and U.S. Semiconductor Technology Levels in the Late 1980s

TECHNOLOGY	JAPANESE LEADERSHIP	EQUAL LEVEL	U.S. LEADERSHIP
Silicon Products			
DRAM	x		
SRAM	x		
EPROM		x	
Microprocessor			x
Custom Products			x
Bipolar	x		
Non-Silicon Products			
Memory	x		
Logic	x		
Linear			x
Optoelectronic	x		
Hetero Structure	x		
Materials			
Silicon	x		
Gallium Arsenide	x		
Processing Technology			
Optical Lithography		x	
Electron Beam Lithography			x
X-ray Lithography		x	
Vapor Deposition		x	
Diffusion		x	
Furnace	x		
Assembling		x	
Packaging	x		
Testing System	x		
CAD		x	
CAM		x	

Source: Department of Defense, U.S. Government, (1987), "Defense semiconductor dependency," quoted in High Technology Study Group (ed.), *Beikoku no gijutsu senryaku* (U.S. technological strategy), (Tokyo: Nikkei Science, 1988), 35.

shares enormously and quite rapidly, dominating the majority shares in eight markets out of eleven in 1988 (Press Journal, 1990:77). Due to needs for maintaining extremely high quality, low prices, and punctual deliveries, semi-conductor companies and materials producers often developed cooperative relationships, and the market tended to be less competitive, though not oligopolistic.

Rapid changes also took place in the semiconductor equipment market. According to Table 2.5 which ranks the top ten semiconductor equipment manufacturers, U.S. companies dominated all ten positions in 1979, with Fairchild Test Systems, Perkin Elmer, and Applied Materials in the top three. However, in 1982, Japanese companies, Advantest, Canon, and Nikon, moved into the fourth, eighth, and tenth places respectively. In 1988, Nikon, Tokyo Electron, and Advantest occupied the top three positions. And in 1989, four out of top five positions and five out of top ten were held by Japanese com-panies. In terms of 1989 sales volume, the five Japanese companies totaled $2,213.5 million, while the five U.S. companies totaled $1,598.7 million, indicating that Japan controlled a significantly higher share. Although Japa-nese equipment manufacturers did not advance as significantly as materials manufacturers, they also came to play a leading position in the world.

Oligopoly has characterized semiconductor equipment markets in Japan. In 1989 two companies dominated in the diffusion furnace (81%), lamp annealer (86%), lithography (85%), photoresist (79%), dicing (88%), and wafer prober (90%) equipment markets. Three companies dominated in the medium current ion implantation (100%) and wafer prober (99%) equip-ment markets, while four companies were dominant in the high current ion implantation (98%), epitaxial (100%), sputtering (95%), and handler (91%) equipment markets. Even the two most dispersed markets, chemical vapor deposition (CVD) and dry etching equipment, had only eight and six compa-nies respectively. Although the tester and bonding markets had eight and six companies respectively, the top company in the tester market dominated 43% and the leader in bonding held 49%, signifying the condition of domination. Foreign companies' shares, including joint ventures with Japanese partners, were high in the high current ion implantation (88%), lamp annealer (81%), epitaxial (66%), CVD (54%), medium current ion implantation (43%), and dry etching (22%) equipment markets (Press Journal, 1991).

The dynamic development of the industry also inspired many new com-panies to enter the field, including MINEBEA in 1984, Kanebo, Kawasaki Steel, Nippon Steel, Tamura, and Toyota in 1985, Kyocera in 1986, Seiko Instruments, Tokairika, and Nissan in 1987, ASCII, Nippon Steel, NKK, and TDK in 1989, and Asahi Glass in 1990. Two new foreign companies also formed joint ventures (Siemens with Fuji Electric, and Westinghouse and General Electric with Mitsubishi), and one foreign company opened a design center (Harris Corporation with the technological help from Toshiba).

An incident that happened in 1984 suggests the severity of domestic competition and investment. In March 1984, when semiconductors were in short supply, NEC, Hitachi, and Matsushita announced plant and equipment investments of ¥110 billion, ¥110 billion, and ¥70 billion respectively. However, when Fujitsu was mistakenly reported to have expanded its investment from the previous estimate of ¥100 billion to ¥125 billion, these companies immediately responded by increasing their planned investment amounts to ¥125 billion, ¥120 billion, and ¥110 billion respectively. Similarly, in April 1984, when NEC announced at a sales representatives' meeting that it would increase its annual sales of semiconductors by 30% to reach ¥495 billion, Hitachi immediately responded by announcing at a stockholders' meeting that it would set a target of ¥520 billion. Naturally, NEC responded to this by announcing its revised target of ¥550 billion (Nikkei Business, 1984a:30).

Thus, the development of the Japanese semiconductor industry is largely indebted to severe competition in domestic product markets and Japanese companies' concentration on commercialized goods and on technological

Figure 2.4 Development of Japanese Semiconductor Industry and Degrees of Innovativeness

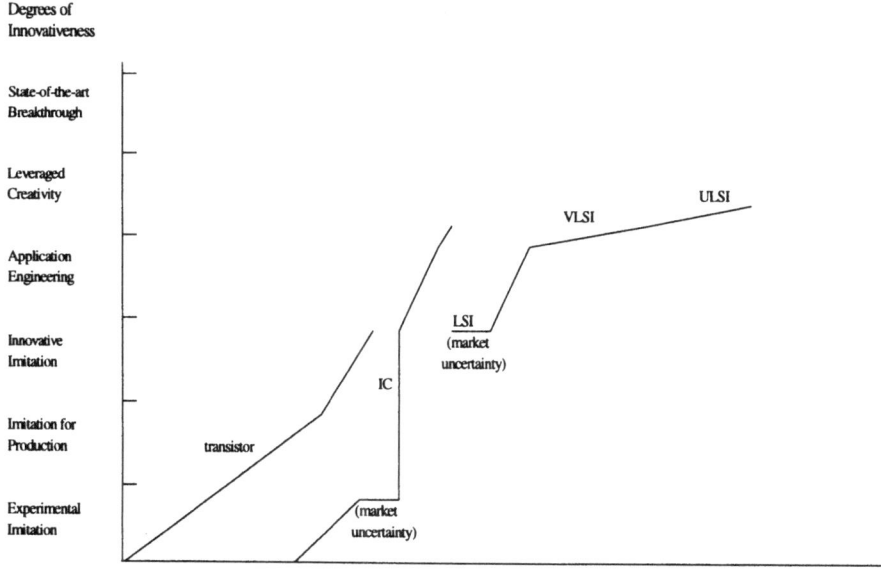

development. The interaction between product market competition and corporate strategies generated transformations in R&D philosophy, which shifted from quick commercialization to market-driven innovation and then to technology-driven innovation. Technological development, market competition, and corporate strategies are closely interrelated (see Figure 2.4), and the severity of product market competition has clearly been the underlying current of the Japanese semiconductor industry. In contrast with the severity of the semiconductor market, equipment markets are oligopolistic, while parts and materials markets are less competitive. Less competitive conditions of equipment and parts/materials markets facilitated the effective development of CCC governance, which became an inseparable part of winning severe competition in the semiconductor market.

Chapter Three

Basic Characteristics of Intra- and Interfirm Interaction

Technological development, competition in the semiconductor market, and corporate strategy interacted in a complex way to generate the dynamics of the Japanese semiconductor industry. To keep winning in an environment of severe market competition, Japanese semiconductor manufacturers constantly searched for ways to cut production costs, automate production, improve product quality, and make technological and product innovations. To ensure their survival, they extended their efforts at making changes to address their relationships with parts, material, and equipment manufacturers and even customers. The outcome of such efforts is the development of competitive-cum-cooperative (CCC) interaction.

In the first stages of this study, I argued that Japanese semiconductor companies relied heavily on CCC governance rather than hierarchical governance. Sales activities and parts, materials, and equipment procurement, all activities which involve frequent human interaction, would rely on CCC interaction, while high asset specificity and needs for maintaining corporate secrets in R&D would compel companies to use internalized interaction for research endeavors. Furthermore, it was predicted that CCC interaction would show an intermediate level of cooperation and interdependence, so that both cooperation-promoting and competition-generating measures could be implemented. But even within CCC interaction, procurement relationships would tend to develop cooperation-oriented CCC interaction, while sales relationship would tend to develop a competition-oriented one.

In this chapter, I provide an estimate of the prevalence of CCC interaction in four functional areas: (1) procurement of parts and materials, (2) procurement of manufacturing equipment, (3) sales, and (4) R&D. Secondly, I examine how the nature and characteristics of CCC interactions differ from the nature of spot and internalized interactions in each functional area. Thirdly, I examine how each company's choice of cooperation-oriented or competition-oriented CCC relations differs with different partners and functional areas.

The findings of this chapter are derived from three sources: (1) a questionnaire survey, (2) interviews, and (3) literature. The survey, which involves a comparison of four different types of partners in four different functional areas, consisted of four types of questionnaires, one for each functional area, and each questionnaire was answered by a manager in charge of the area in question. In each questionnaire (i.e. each functional area), respondents were asked to answer the same set of questions about each type of partner. This means that each company answered a similar set of questions sixteen times, which helped to provide distinctions between the characteristics of diverse partners as well as of different functional areas.

Each questionnaire described four different types of partners, who were involved in vertical, horizontal, spot, and internalized interactions, in that order, as follows:

(1) companies with a weaker overall capability than yours, with which you have a long-term relationship (e.g. closely- and vertically-affiliated companies [vertical *keiretsu*], small- and medium-sized companies, and venture firms);

(2) companies with a similar overall capability as your company, with which you have a long-term relationship (e.g. closely- and horizontally-affiliated companies [horizontal *keiretsu*], independent companies, and alliance partners);

(3) companies in temporary relationships (e.g. spot markets and arm's length transactions);

(4) intrafirm divisions and sections, and majority-owned subsidiaries and foreign affiliates

Respondents answered all questions, keeping these descriptions in mind.

A.   Prevalence of Competitive-cum-cooperative Interaction in Business Transactions

How prevalent is CCC interfirm interaction in the Japanese semiconductor industry? In the survey, respondents were asked to identify the proportion of business transactions conducted with different types of partners in each functional area.[1]

On average, semiconductor manufacturers purchased approximately 47.9% of the total transacted value of parts and materials from horizontally-related companies (horizontal CCC interaction, see Table 3.1), and 33.1% from vertically-related companies (vertical CCC interaction). Intrafirm and interaffiliate procurement of parts and materials represented approximately 10.7% of purchases (internalized interaction), while parts and material purchases from companies in the spot market were estimated at 8.3% (spot interaction). These figures clearly show that in the procurement of parts and materials, horizontally- and vertically-related companies are two major partners for semiconductor manufacturers, covering 81.0%.

Semiconductor manufacturers purchased, on an average, 48.4% of the total initial equipment investment from horizontally-related companies, 25.5% from vertically-related companies, 13.2% through intrafirm procurements, and 12.9% from companies in the spot market. In this area, CCC interaction is again predominant, covering 73.9% of manufacturing equipment purchases.

The slightly higher figure found for purchases in the spot market requires an explanation. Machines developed for a specific product inside a leading semiconductor manufacturer were usually used intensively for about four to five years, which was too short for recovering the enormous costs required for their development. Hence, semiconductor manufacturers started selling machinery two years after the beginning of mass production to any buyer in the market, despite the firm-specific nature of their technology. Some of foreign and second-tier Japanese semiconductor manufacturers became regular customers. Also, when other leading semiconductor manufacturers needed to expand their production facilities, they purchased some inexpensive equipment in the spot market.[2]

In sales, horizontally-related customers, on average, purchased about 50.6% of semiconductor companies' total sales values. Although vertically-

---

[1] Each questionnaire asked respondents to specify the estimated proportions of transactions conducted with each type of company. For example, for procurement of parts and materials, the questions were as follows:

    (1) What percentage of the total cost of the procurement of parts and materials is used for purchasing from companies with which you have a long-term relationship and which have a weaker overall capability than your company, (e.g. closely- and vertically-affiliated companies [vertical *keiretsu*], small- and medium-sized companies and venture firms)?;

    (2) What percentage of the total cost of the procurement of parts and materials is used for purchasing from companies with which you have a long-term relationship and which have a similar overall capability as your company, (e.g. closely- and horizontally-affiliated companies [horizontal *keiretsu*], independent companies and alliance partners)?;

    (3) What percentage of the total cost of the procurement of parts and materials is used for purchasing from companies in a temporary relationship with your company (e.g. spot markets and arm's-length transactions)?; and

    (4) What percentage of the total cost of the procurement of parts and materials is used for internal purchasing (intrafirm divisions and sections, and majority-owned subsidiaries and foreign affiliates)?

A similar format was used for the other functional areas.

[2] Interview with Company Q manager.

Table 3.1 Estimated Percentage of Total Transacted Value By Type of Interaction and Functional Area

	FUNCTIONAL AREAS			
TYPE OF INTERACTION	PROCUREMENT OF PARTS AND MATERIALS (N = 5)	PROCUREMENT OF MANUFACTURING EQUIPMENT (N = 5)	SALES (N = 6)	R&D (N = 5)
Spot Interaction	8.3%	12.9%	8.2%	0.6%
Horizontal CCC Interaction	47.9	48.4	50.6	5.4
Vertical CCC Interaction	33.1	25.5	22.2	2.2
Internalized Interaction	10.7	13.2	19.0	91.8
Total	100.0	100.0	100.0	100.0

related companies were still the second most important customers, this type of buyer occupied a less important position (22.2%) than procurement. The major reason for this smaller percentage is the increased importance of intrafirm sales, which was estimated to cover, on an average, 19.0% (internalized interaction). Intrafirm sales mostly involve transactions between semiconductor manufacturing divisions and other divisions producing such electronics goods as video players and televisions. Sales to companies in the spot market (8.2%) were the least important. CCC interaction (72.8%) remains as the most important sales relationship.

Horizontally- and vertically-related companies (5.4% and 2.2% respectively) were not very important partners for R&D. Intrafirm expenditure covered, on an average, approximately 91.8% of the total semiconductor-related R&D budget. A portion used for one-time purchases of technology or R&D-related services was estimated at only 0.6% (spot interaction). Even the amount of R&D budget used for alliances seems to have been insignificant.

Hence, CCC interaction, which involves both vertically- and horizontally-related companies, covered more than 70% of the total transacted value in all functional areas except R&D. These functional areas extensively involve human relations, where, in the Japanese context, the institutional inheritance of interpersonalism seems to have created an environment conducive to both cooperation-promoting and competition-generating measures. In contrast, the high asset specificity and need for maintaining corporate secrets that are characteristic of an R&D environment seem to have compelled companies to use internalized interaction. These findings support the arguments presented earlier.

B. Nature and Characteristics of Interaction

Within a short period of time, the Japanese semiconductor industry came to occupy a world-leadership position. As the previous section indicates, CCC interaction has been very important in generating the industrial dynamics of all functional areas except R&D. What are the nature and characteristics of interaction with each type of partner in each functional area? In Chapter One, several theories on governance indicated that the way interacting partners are harmonized is greatly influenced by cooperation, interdependence, power relationships, frequency of interaction, asset specificity of products, and shared goals and interests. In this chapter, these factors are examined to identify the nature and characteristics of intra- and interfirm interaction; the factor of shared goals and interests, however, is not explored here but will be explored in Chapters Four and Five.[3]

---

[3] Respondents were asked to select the most important company in each category of partner and to answer questions about that company. For example, to acquire information about the characteristics of vertical CCC interaction in the procurement of parts and materials, questions were written as follows:

Please select the most important parts and materials manufacturer that has the following characteristics: a weaker overall capability than your company and a long-term relationship with your company (e.g. closely- and vertically-affiliated companies [vertical *keiretsu*], small- and medium-sized companies and venture firms). And please answer the following questions about the company:

(1) Characterize the relationship between your company and the company you have identified.
  (a) Degree of interdependence
      (1. very weak, 2. weak, 3. intermediate, 4. strong, 5. very strong)
  (b) Degree of cooperation
      (1. very weak, 2. weak, 3. intermediate, 4. strong, 5. very strong)
  (c) Frequency of interaction (choose a category and specify the number of times)
      ( ) times/day   ( ) times/week   ( ) times/month   ( ) times/year
  (d) Degree of firm (asset) specificity of products
      (1. products with no firm-specific characteristics,
        2. products with a low degree of firm-specific characteristics,
        3. products with an intermediate degree of firm-specific characteristics,
        4. products with a high degree of firm-specific characteristics,
        5. products with a very high degree of firm-specific characteristics.)
  (e) What are the goals and interests that your company shares with the selected company?
      (   )
  (f) Degree of capability to coerce the company
      (1. very weak, 2. weak, 3. intermediate, 4. strong, 5. very strong)
  (g) Degree of vulnerability to be coerced by the company
      (1. very weak, 2. weak, 3. intermediate, 4. strong, 5. very strong)

This survey had a sample size of only seven, which makes any sophisti-cated statistical analysis impossible and meaningless. Hence, comparisons of means are considered the most robust way to interpret the data, and mean values were classified in each type of interaction into eight categories as follows: weak (mean value = 1.5 - 1.9); weak+ (2.0 - 2.4); medium (2.5 - 2.9); medium+ (3.0 - 3.4); strong (3.5 - 3.9); strong+ (4.0 - 4.4); and very strong (4.5 - 5.0). When two mean values had a difference of less than 0.2, they were interpreted as having no difference. For example, when one score was 3.20, while the other score was 3.39, they were treated as if they were very similar.

1.   Procurement of Parts and Materials

Semiconductors are highly sensitive to slight changes in machinery, factory environment, and quality of parts and materials. When air or water is not clean enough, when machinery is not operating stably, or when the quality of material is not pure enough, these conditions immediately result in a lower cropping rate of semiconductors. To ensure the high quality of parts and materials, semiconductor companies and suppliers use computer-integrated operations and obtain detailed records of each delivered lot. When a prob-lem occurs, both companies cooperate intensively to identify the source of problems. The nature of this operation suggests that cooperation is a very important part of making sure semiconductor production works smoothly.

According to Table 3.2, internalized interaction between semiconductor divisions and intrafirm parts/material supply units scored highest in degree of interdependence (a strong+ level) and cooperation (a very strong level). In contrast, semiconductor companies' interactions with suppliers in the spot market show the lowest scores (both at a weak+ level). Clearly, intrafirm procurement involves a strong sense of cooperation and interdependence, while interaction with suppliers in the spot market can hardly nurture these qualities. Between the two extremes lies CCC interaction, and within the category of CCC interaction, there was a large difference between vertical and horizontal interaction. Vertically-related suppliers were highly coopera-tive and interdependent with semiconductor companies (at a strong+ level), while horizontally-related suppliers were less so (at a medium+ level). For CCC interaction to be highly effective, the intermediate level of cooperation and interdependence becomes quite important. If cooperation is too strong, the implementation of competition-generating measures becomes difficult. If it is too weak, companies will also be unable to implement competition-generating measures, because many of these measures require a good coop-erative environment.

Table 3.2 Nature and Characteristics of Interaction in Procurement of Parts and Materials

CHARACTERISTICS	TYPE OF INTERACTION			
	SPOT INTERACTION (N = 6)	HORIZONTAL INTERACTION (N = 5)	VERTICAL INTERACTION (N = 6)	INTERNALIZED INTERACTION (N = 6)
Degree of Interdependence*	2.00	3.40	4.00	4.40
Degree of Cooperation*	2.17	3.40	4.00	4.60
Degree of Capability to Coerce*	2.33	3.20	3.67	4.00
Degree of Vulnerability to be Coerced*	2.33	2.60	2.80	2.40
Degree of Asset Specificity of Products*	1.89	3.40	3.22	3.13
Frequency of Interaction (Times/Year)	24.3	96.0	105.6	154.8

Note: *1.0 means very weak and 5.0 means very strong. Scores are categorized as follows: very weak (mean value = 1.0 - 1.4); weak (1.5 - 1.9); weak+ (2.0 - 2.4); medium (2.5 - 2.9); medium+ (3.0 - 3.4); strong (3.5 - 3.9); strong+ (4.0 - 4.4); and very strong (4.5 - 5.0).

Similarly, intrafirm procurement tended to involve more frequent interaction than procurement from other types of partners, while semiconductor companies interacted least frequently with suppliers in the spot market. Vertically- and horizontally-related suppliers interacted at a level between the two extremes of intrafirm and market procurement. Again, CCC interaction was found to have an intermediate level of frequency in interaction.

It is, however, a little strange that while the highest percentage of parts and materials were purchased from horizontally-related suppliers (48.38%), intrafirm procurement involved the highest rate of interaction. Semiconductor divisions usually purchased parts and materials from intrafirm parts/materials supply units, which made bulk purchases of parts and materials by coordinating the demands of diverse divisions within the same company. The purchasing units often became the center of information about standardized parts and materials. Consequently, interaction with these units became more frequent.[4]

Using asset-specific products means that particular characteristics of prod-

---

[4] Interview with Company Q manager.

ucts link transacting partners together, making it harder for companies to change partners. All types of partners, except suppliers in the spot market, sold parts and materials with a medium+ level of asset specificity. Spot market suppliers had an extremely low capability for handling asset-specific parts and materials, and this capability is one indispensable condition for CCC and internalized interaction.

Power relationships differentiate the characteristics of interaction. The difference between the two items of Degree of Capability to Coerce and Degree of Vulnerability to be Coerced was calculated to find the Degree of Mutual Influence. This score was then categorized from a semiconductor company's perspective as follows:

(1) suppressive control (1.2 ≤ score)

(2) favorable mutual influence (0.2 ≤ score < 1.2)

(3) symmetric mutual influence (-0.2 ≤score < 0.2)

(4) unfavorable mutual influence (-1.2 ≤score < -0.2)

Suppressive control was defined as a case in which a semiconductor company had a strong capability to influence a partner's behavior, while the partner did not have the power to influence the semiconductor company. Favorable mutual influence was identified as case in which both the semiconductor company and its partner influenced each other, but the former could use its superior power position to influence its partner in its own favor. Symmetric mutual influence existed when both partners influenced each other equally. And unfavorable mutual influence was defined as a case in which the semiconductor company and its partner influenced each other, but the partner used its superior power position to achieve outcomes unfavorable to the semiconductor manufacturer.

Semiconductor divisions reported having a strong capability to coerce intrafirm supply units (at a strong+ level), and they were hardly vulnerable to coercion (at a weak+ level). The difference between the scores was 1.60, signifying suppressive control. Since intrafirm supply units are bulk-purchasing agents that support the semiconductor divisions' operations, they can hardly be in a position to influence the operations of the divisions. The relationship of suppressive control with a very strong sense of cooperation that was found to govern these interactions is characteristic of bureaucratic mechanisms and hierarchical governance. These structures are often found when intrafirm relations are developed between major and minor supporting units. The opposite case was found in spot interactions, in which semiconductor companies neither coerced suppliers in the spot market (at a weak+ level) nor were coerced by them (at a weak+ level). They were independent of each other without any significant mutual influence (the difference of scores is 0.0).

Mutual influence was found to be a distinctive characteristic of CCC interaction. Semiconductor manufacturers strongly coerced vertically-related suppliers (at a strong level), while the latter also influenced the former (at a medium level). Similarly, semiconductor companies coerced horizontally-related suppliers (at medium+ level), but they were also vulnerable to be coerced by the suppliers (at a medium level). Score differences are 0.87 for vertically-related suppliers and 0.60 for horizontally-related suppliers, signifying a condition of favorable mutual influence for semiconductor manufacturers. CCC interaction developed a ground for mutual influence, though asymmetry did exist in the power relationships in favor of semiconductor companies. Vertically-related suppliers had the highest capability to influence; this was indicated by the degree of semiconductor companies' vulnerability to be coerced. Vertically-related suppliers seem to interact dynamically with semiconductor companies and participate in some decision-making.

There are two interesting findings with regard to mutual influence. One is that semiconductor manufacturers usually had a stronger power position vis-á-vis parts and material suppliers, except with those in the spot market. Even large-sized horizontally-related companies were vulnerable to semiconductor manufacturers. This is a distinctive characteristic found only in the procurement of parts and materials. As discussed in Chapter Two, semiconductor companies faced severe market competition. They needed to outperform competitors in the price and quality of semiconductors by cutting procurement costs and ensuring high quality in parts and materials. Hence, they pressured parts and material producers much more than they did to equipment manufacturers. The other finding is that mutual influence, the characteristic of CCC interaction, was not found in internalized or spot interaction.

Thus, in the procurement of parts and materials, contrasts among different types of interaction were rather clear. Semiconductor divisions strongly cooperated with intrafirm supply units to purchase asset-specific products. Their relationship, however, tended to be suppressive in favor of the semiconductor divisions, and lacked an environment of mutual influence. Apparently, suppressive control with a strong sense of cooperation is a way bureaucratic mechanisms harmonize major and minor intrafirm actors under hierarchical governance. Companies in the market marginally cooperated with semiconductor manufacturers, and remained independent without any mutual influence. The lack of cooperation between them has made it impossible to handle any asset-specific products. Without a doubt, market governance is harmonizing their interactive behavior.

Between the two extremes lies CCC interaction. Vertically- and horizontally-related companies cooperated with semiconductor manufacturers at an intermediate level, and supplied some asset-specific products. This intermediate level of cooperation became an important condition for introducing competition-generating measures. Though power relationships between these suppliers and semiconductor manufacturers were asymmetric in favor of the latter, their relationships were mutually influential. This is a distinct characteristic of CCC interaction. Vertically-related suppliers showed the highest capability to influence, despite the power asymmetric relations. Hence, some of CCC interaction's unique characteristics for these companies were an intermediate level of cooperation and interdependence, mutual influence, and an intermediate level of involvement of asset-specific products. Because of the importance of maintaining quality in parts and materials and the above characteristics, partners in CCC interaction seemed to be harmonized under cooperation-oriented CCC governance. Chapter Four provides further evidence to support this point.

2.   Procurement of Manufacturing Equipment

Unlike providing parts and materials, semiconductor manufacturing equipment requires more sophisticated technology, and this equipment is purchased from the oligopolistic market. Manufacturing equipment is firm-specific, and in many cases semiconductor and equipment manufacturing companies jointly develop it. Once manufacturing equipment is produced in cooperation with a manufacturer, that manufacturer's technology confines and restricts choices for new developments. Hence, technology itself becomes a source of cooperation and long-term relations.

According to Table 3.3, intrafirm procurement showed very strong levels of cooperation and interdependence. In contrast, semiconductor manufacturers only marginally cooperated or became interdependent with companies in the spot market (at weak and weak+ levels). Vertically- and horizontally-related equipment suppliers stood between the two extremes. Vertically-related suppliers showed a strong sense of interdependence and cooperation with semiconductor manufacturers, while horizontally-related suppliers indicated a medium level. All partners, except companies in the spot market, transacted with an intermediate level of asset specificity (at a medium+ level). This had much to do with the firm-specific nature of equipment jointly pro-

Table 3.3 Nature and Characteristics of Interaction in Procurement of Manufacturing Equipment

CHARACTERISTIC	TYPE OF INTERACTION			
	SPOT INTERACTION (N = 6)	HORIZONTAL INTERACTION (N = 5)	VERTICAL INTERACTION (N = 5)	INTERNALIZED INTERACTION (N = 5)
Degree of Interdependence*	1.60	2.80	4.00	4.50
Degree of Cooperation*	2.00	2.80	3.80	4.50
Degree of Capability to Coerce*	2.00	2.80	3.60	3.25
Degree of Vulnerability to be Coerced*	2.20	2.80	2.40	3.00
Degree of Asset Specificity of Products*	2.33	3.40	3.40	3.00
Frequency of Interaction (Times/Year)	7.0	91.8	85.4	42.0

Note: *1.0 means very weak and 5.0 means very strong. Scores are categorized as follows: very weak (mean value = 1.0 - 1.4); weak (1.5 - 1.9); weak+ (2.0 - 2.4); medium (2.5 - 2.9); medium+ (3.0 - 3.4); strong (3.5 - 3.9); strong+ (4.0 - 4.4); and very strong (4.5 - 5.0).

duced, while suppliers in the spot market tended to deal with standardized products. These findings are quite similar to the findings regarding the procurement of parts and materials. But a very different situation existed in terms of power relationships.

Manufacturing equipment involves highly sophisticated technology, and the joint development of firm-specific equipment tends to lock interfirm relations. Because of these characteristic, semiconductor companies did not stand in a favorable power position as they did in relation to parts and materials suppliers. This was especially true with horizontally-related companies and intrafirm equipment manufacturing units.

Horizontally-related equipment manufacturers are large-sized Japanese and American companies, that dominate markets oligopolistically, for example, Nikon, Advantest, Tokyo Electron, TELVARIAN (US-Japan joint venture), and Applied Materials Japan (US company). In the late 1980s, Nikon monopolized 62% of the lithography equipment market, while Advantest handled 43% and 45% of the tester and handler markets respectively (Press Journal, 1991). Since, on an average, 48.4% of business transactions were made with horizontally-related equipment manufacturers, semiconductor companies interacted with them most frequently. Naturally, they interacted with these types of companies less frequently than with horizontally-related parts and materials suppliers, simply because procuring manufacturing equipment took place relatively infrequently. These partners equally influenced each other (at a medium level with a score difference of 0.0). Thus, these relations can be categorized as symmetric mutual influence.

Within their relationships, semiconductor divisions had slightly more power than intrafirm equipment manufacturing units (both at a medium+ levels with a score difference of 0.25), but these partners mutually influenced each other. Unlike the case of intrafirm parts/material procurement, that involved suppressive control, this relationship can be categorized as mutual influence that is a bit favorable for semiconductor divisions (favorable mutual influence). The intrafirm units engaged in developing frontier technology that required intensive cooperation and mutual influence with semiconductor divisions. This characteristic resembles the internalized interaction in R&D (presented later in this chapter). Thus, interaction for R&D purposes seems to involve mutual influence suitably harmonized under hierarchical governance.

Despite a strong sense of cooperation and interdependence, semiconductor manufacturers exercised suppressive control over vertically-related equipment manufacturers. The former strongly coerced the latter, while the latter hardly had any capability to coerce the former (at the level of weak+ with a score difference of 1.20). The power asymmetry was quite big, probably because vertically-related suppliers, usually small- and medium-sized companies, grew dependent on particular semiconductor manufacturers. Semiconductor manufacturers interacted with these companies as frequently as they worked with horizontally-related suppliers, and their rate of interaction was higher than with other types of partners. These companies usually specialized in less sophisticated equipment, such as testing equipment. But was the interaction between semiconductor manufacturers and vertically-related companies really cooperation-oriented CCC interaction? Was the influence of vertically-related companies too weak to call the relationship one of mutual influence? The answers to these questions become clear in the next chapter.

Suppliers in the spot market showed a weak capability to coerce and weak vulnerability to being coerced (both at weak+ level), indicating that hardly any mutual influence took place.

Thus, the procurement of manufacturing equipment had a tendency in cooperation, interdependence, and asset-specificity that was similar to parts and materials. Companies in the spot market behaved similarly to those in the spot market for parts and materials, suggesting the influence of market governance. Intrafirm procurement was based on a very strong sense of cooperation and favorable mutual influence. This characteristic was quite different from those of parts and materials interaction, within which semiconductor divisions had a high level of suppressive control over parts/material supply units. Since intrafirm equipment manufacturing units engaged in R&D for equipment development, interaction seems to have been harmonized in a way favorable for R&D under hierarchical governance. Chapter Four provides further evidence to support this point.

CCC interaction in the procurement of manufacturing equipment did not involve many dynamic interactions, despite the fact that it covered 73.9% of transactions. Horizontally-related companies interacted frequently, but with a weaker sense of cooperation and interdependence than vertically-related companies, while the latter cooperated well, but under the suppressive control of the semiconductor companies. Owing to their oligopolistic positions, technological capability, and low need for continuous interaction (other than maintenance), horizontally-related suppliers developed an intermediate level of cooperation and interdependence with semiconductor manufacturers. Both partners mutually influenced each other at a medium level, and established genuine power-symmetry. Given these characteristics, horizontally-related suppliers seem to have been interacting with semiconductor companies under the harmonizing mechanism of cooperation-oriented CCC governance. However their medium level of cooperation and interdependence makes horizontally-related companies less conspicuous actors in CCC governance. In contrast, semiconductor companies exercised suppressive control over vertically-related suppliers, even though the latter had interactions as frequently as horizontally-related suppliers. The suppressive control of semiconductor manufacturers over vertically-related suppliers makes it doubtful whether cooperation-oriented CCC governance was the harmonizing mechanism. CCC interaction in the procurement of manufacturing equipment requires more detailed investigation in Chapter Four.

3.   Sales

Long-term relationships in sales were not at all similar to those in parts, materials, and manufacturing equipment procurement. Interviews with seven sales managers (Companies P, Q, R, T, U, V, and X) revealed that the relationship was simply a series of consecutive short-term contracts won by salespersons, which made it look long term. Developing a long-term relationship would be ideal for a semiconductor manufacturer, since these relationships provide stable purchases and prices, which could constitute a good ground for planning in production and R&D.[5] But severe market competition among semiconductor manufacturers has prevented such development.

The lack of genuine long-term relationships does not, however, mean that CCC interaction does not take place. Intensive competition has increasingly compelled salespersons, system engineers, or applied engineers, (whichever name they are called by each company), to establish business contacts in product development departments rather than in procurement ones. Obtaining information from procurement departments would make the salesperson too late to win in competition. In the 1970s, 70 - 80% of salespersons' visits were to procurement departments and only 20 - 30% were to product development departments. But in 1992 about 50% were to the latter, and this percentage was expected to increase to about 70%.[6] This shift was partly due to the increasing usage of application specific integrated circuits (ASICs), which required intensive discussions and information exchange with product development engineers. Once design is completed, ASICs are also purchased by procurement departments.[7] Even business dealings on standardized integrated circuits (ICs) were sometimes discussed at the development stage, although in many cases, as with ASICs, procurement departments handled purchases.

Product development departments became more important contacting places, because obtaining information on who is developing and designing which products as well as learning who is in charge of a project has become an important aspect of sales activities. Once information has been obtained, then a salesperson could aggressively make suggestions and proposals, and even provide some blue prints. How fast a salesperson could get involved has determined whether he or she was one step ahead of competitors. But to obtain such information and conclude the first contract has been a tough procedure, and has required frequent contacts.

After successful completion of several contracts, a customer can come to develop a sense of trust and reliability with a semiconductor manufacturer. Then, establishing this type of relationship can increase the probability of obtaining information about the next project before competitors can do so. As a matter of fact, in 1980 51% of integrated circuit projects were made by the initiative of semiconductor manufacturers rather than buyers (Kikai Shinko Kyokai Keizai Kenkyusho, et al., 1980:67). This is why semiconductor com-

---

[5] Interviews with managers in Companies Q and R.

[6] Interview with Company Q manager.

[7] Interviews with managers in Companies Q, R, T, U, and V.

panies have interacted more frequently in sales activities than any other functional areas. (The averages of all types of partners were 223 times per year for sales, 95 for the procurement of parts and materials, 64 for R&D, and 57 for the procurement of manufacturing equipment). The need for frequent interaction has also compelled many semiconductor manufacturers to sell their products through sole agents, which generally increases channels of contacts to customers. Companies Q, R, U, V, and X made no less than 50% of their sales through sole agents in 1992, while Companies S and T were dependent more on direct sales.

Sole agents usually maintain good relationships with engineers in customers' product development and purchasing departments. And when information about a new product is obtained, engineers from sole agents and semiconductor manufacturers work together closely to develop suggestions and proposals to customers.[8]

Engineers in the product development department of a large customer company, however, usually move from the department to a factory to engage in manufacture of their new products, and their success in those products very much determines their future promotion. Hence, they tend to be very eager to develop the best product and generally are highly sensitive about diverse aspects of semiconductors, including the functions, quality, price, and failure rates. These engineers often deliberately release only the minimum amount of information in order to exploit competition among semiconductor manufacturers and obtain new ideas and suggestions at the designing stage.[9] This process has been especially prevalent in ASICs, because competition in a standard product like a DRAM IC takes place over samples. In the companies involved in this study, interfirm sales relationships were never locked in for the long term; they lasted through the duration of short term contracts. This explains why all types of interaction had a relatively low level of asset specificity in their products (at weak and weak+ levels in Table 3.4).

Severe competition like this clearly impacts semiconductor companies' relations with horizontally- and vertically-related customers. The companies discussed here developed an intermediate level of cooperation and interdependence with these types of customers (at a medium+ level). But they became quite vulnerable to horizontally-related customers (unfavorable mutual influence with a score difference of -0.28), since horizontally-related

---

[8] Interviews with managers in Companies Q, R, and U.
[9] Interview with Company R manager.

Table 3.4 Nature and Characteristics of Interaction in Sales

CHARACTERISTIC	TYPE OF INTERACTION			
	SPOT INTERACTION (N = 7)	HORIZONTAL INTERACTION (N = 7)	VERTICAL INTERACTION (N = 7)	INTERNALIZED INTERACTION (N = 6)
Degree of Interdependence*	1.57	3.14	3.14	4.33
Degree of Cooperation*	1.71	3.29	3.29	3.83
Degree of Capability to Coerce*	3.00	2.43	2.86	3.33
Degree of Vulnerability to be Coerced*	1.71	2.71	2.71	2.83
Degree of Asset Specificity of Products*	1.57	2.11	1.95	2.33
Frequency of Interaction (Times/Year)	21.1	183.7	287.1	400.5

Note: *1.0 means very weak and 5.0 means very strong. Scores are categorized as follows: very weak (mean value
= 1.0 - 1.4); weak (1.5 - 1.9); weak+ (2.0 - 2.4); medium (2.5 - 2.9); medium+ (3.0 - 3.4); strong (3.5 - 3.9);
strong+ (4.0 - 4.4); and very strong (4.5 - 5.0).

customers were procuring large quantities of semiconductors, were developing forefront technologies, and were stringent regarding technology and performances. Horizontally-related customers were the only partners that put semiconductor manufacturers in a very vulnerable bargaining position. Even vertically-related customers, which tended to be smaller in size and procure less quantity, could stand in an equal bargaining position vis-á-vis semiconductor manufacturers (symmetric mutual influence with a score difference of 0.15). Again, sales interactions presented the only cases of vertically-related companies standing in an equal bargaining position with semiconductor manufacturers. These types of customers were involved in 72.8% of all sales transactions (Table 3.1). Their relative power in transactions with semiconductor companies suggests that market competition is extremely severe.

These findings strongly suggest that the delicate mixture of cooperation and competition enabled by CCC interaction is quite valuable for winning in market competition. Cooperation is merely the essence of long-term relationships; instead has become a necessary means for obtaining information and winning consecutive contracts. Uncertainty is the basic condition of interaction in sales, and CCC interaction functions to reduce the inherent uncertainty. This kind of vertical and horizontal interaction is called competition-oriented CCC interaction. Consecutive winning of contracts results in a pseudo-long-term relationship. This partnership looks like a long-term relationship, but in reality it consists of consecutive short-term contracts, each of which has little locking effect and faces competition from other manufacturers. Severe product-market competition, salespersons' struggles, and buyers' exploitation of competition provide the foundation for this type of relationship, in which cooperation becomes a means of winning contracts and reducing uncertainty from severe market competition. Buyer-seller interaction is harmonized in this manner under competition-oriented CCC governance.

As was the case with parts and materials and equipment acquisition, internalized sales interactions were characterized by strong cooperation, while spot interaction failed to develop cooperation (at a weak level). Existing between the two extreme cases, CCC interaction showed an intermediate level of cooperation and interdependence (at a medium+ level). Again, this intermediate level of cooperation seems to have provided a favorable environment for developing CCC governance.

Under the severe competition seen by these companies, a strange phenomenon took place. According to Table 3.4, even though horizontally-related companies made most of the purchases, semiconductor divisions interacted with intrafirm customers most frequently. Intrafirm relationships developed a strong sense of interdependence (at a strong+ level), but a little lower level of cooperation (at a strong level). In all other functional areas, they have shown the same level in both indicators of cooperation and interdependence. Furthermore, as in CCC interactions, the semiconductor division and intrafirm customers faced each other under an asymmetrical power relationship (favorable mutual influence at a medium+ level of capability to coerce and a medium level of vulnerability to be coerced with a score difference of 0.5). These findings suggest a condition similar to the vertical CCC interaction found in the procurement of parts and materials. The high frequency of interaction may have been caused by the frequent requests that intrafirm customers made to develop high-risk integrated circuits, which would require intensive interaction. But low asset specificity of products (at a weak+ level) suggests that specialized IC developments were a rather small portion of business. The profit-center orientation of semiconductor divisions and intrafirm customers may have caused them to operate as if they were two independent companies, and their interaction may have created a market-like environment within a bureaucratic organization. Chapter Five provides further evidence for this interpretation.

Interaction between semiconductor manufacturers and customers in the spot market tended to be rather neutral with little sense of cooperation or mutual influence. But the relatively strong power position of semiconductor manufacturers vis-á-vis the customers was due to the 1986 U.S.-Japan semiconductor agreement, which reduced discount sales in gray markets. The reduced supply in the market forced semiconductor traders to negotiate with semiconductor manufacturers for supplies with lower prices, which placed semiconductor traders in a highly vulnerable position. Since it was a temporary phenomenon, the harmonizing mechanism can still be classified as market governance.

Thus, the severe market competition in sales generated a very different type of behavior than was generated by competition in other functional areas. It allowed horizontally- and vertically-related customers to take advantage of the vulnerable position of semiconductor manufacturers and to enforce stringent criteria for selecting a supplier in each contract. To win competition and reduce uncertainty, semiconductor manufacturers developed pseudo-long-term relations by using cooperation as a means to obtain new project information and participate in the planning and designing stages. More specifically, they engaged in competition-oriented CCC interaction, which took place in as much as 72.8% of sales transactions. Intrafirm sales also involved both low asset-specificity products and high-risk ICs. Intensive cooperation and interaction become important for developing high-risk ICs, but for less asset-specific ICs, a market-like environment created by interacting divisions operating as profit-centers may have been at work. Companies in the spot market only marginally generated a sense of interdependence and cooperation.

4.   R&D

According to Table 3.1, intrafirm R&D occupied, on an average, 91.8% of the total semiconductor-related R&D budget. Intrafirm R&D was especially important for developing advanced devices and maintaining product-market leadership by innovation. Achieving these two goals required a complex R&D system.

The leading semiconductor companies had a three-step research structure in place in 1992, and targeted for (1) basic research, (2) applied research, and (3) research for immediate application. The first level was usually carried out at a central research laboratory, where researchers reported directly to top executives; the second level of research was done at laboratories in specific company divisions, and these laboratories were ultimately overseen by the top executive of a group of divisions; and the third level of R&D was done within departments inside of division or factories, which were over-seen by the head of a division or a factory. These three areas of research were expected to produce results over a different time periods: level one was to produce results after three to ten years, level two in one to three years, and level three in about one and a half years.

For example, in NEC basic and device-development research related to the integrated circuit was conducted mainly in the Basic Research Institute and the Microelectronics Research Institute within the Research and Development Group. Some basic R&D was also conducted in the Functional Devices Research Laboratories and the Opto-Electronics Research Laboratories within the same group. Applied research for promoting technological advancement was conducted in the ULSI Device Development Laboratories and the ULSI Systems Development Laboratories within the Semiconductor Group.[10] Finally, design developments to satisfy the immediate needs of customers and technology developments to satisfy production units were conducted in departments within each semiconductor division.[11]

As this structure shows, R&D involved diverse research laboratories, centers, and divisions. Undeniably, effective R&D within a company required very strong interdependence and cooperation, as well as intensive interaction. According to Table 3.5, internalized interaction between semiconductor divisions and intrafirm research-related units developed a very strong sense of cooperation and interdependence, and they interacted frequently. The intensity of cooperation and interaction was necessary not only for research, but also for secrecy maintenance. Interactive behavior and secrecy maintenance cannot be realized without strong organizational identity and hierarchical control.

Both semiconductor divisions and R&D sections influenced each other

---

[10] Information from NEC.

[11] At Toshiba, basic research on the semiconductor was carried out in the ULSI Research Center which reported directly to top executives. Applied research, such as developing engineering samples and mass production technology, was conducted in the Semiconductor Device Engineering Laboratory and the Engineering Department of the Semiconductor Group, both of which reported to the head of the Group. Engineering departments in factories developed commercial samples and improved production technology (information from Toshiba).

Similarly, at Hitachi, basic research was conducted in the Advanced Research Laboratory, the Central Research Laboratory, the Hitachi Research Laboratory, the Mechanical Engineering Research Laboratory, the Production Engineering Research Laboratory, and the Microelectronics Products Development Laboratory. Applied research was conducted in the Device Center, and R&D for immediate application was conducted in departments within factories (information from Hitachi).

Table 3.5 Nature and Characteristics of Interaction in R&D

CHARACTERISTIC	TYPE OF INTERACTION			
	SPOT INTERACTION (N = 6)	HORIZONTAL INTERACTION (N = 5)	VERTICAL INTERACTION (N = 6)	INTERNALIZED INTERACTION (N = 6)
Degree of Interdependence*	1.50	3.17	3.67	4.67
Degree of Cooperation*	2.00	3.83	3.67	4.67
Degree of Capability to Coerce*	2.50	2.17	4.33	3.67
Degree of Vulnerability to be Coerced*	2.50	2.17	2.33	2.83
Frequency of of Interaction (Times/Year)	14.0	16.8	40.0	184.2

Note: *1.0 means very weak and 5.0 means very strong. Scores are categorized as follows: very weak (mean value = 1.0 - 1.4); weak (1.5 - 1.9); weak+ (2.0 - 2.4); medium (2.5 - 2.9); medium+ (3.0 - 3.4); strong (3.5 - 3.9); strong+ (4.0 - 4.4); and very strong (4.5 - 5.0).

with the former having a slightly stronger power position (at a strong level of capability to coerce and a medium level of vulnerability to be coerced with a score difference of 0.84). Their interaction showed the characteristics of favorable mutual influence, suggesting that R&D required mutual influence more than control. Hence, favorable mutual influence with strong interdependence and cooperation can be identified as the characteristics of hierarchical governance in R&D related operations.

Semiconductor manufacturers and horizontally-related companies interacted in order to achieve product-market leadership through R&D. But R&D constituted only 5.4% of the semiconductor companies' budgets. In many cases, companies formed alliances with foreign or domestic companies for technological innovation. Their interaction, however, took quite a different form than internalized interaction did. Horizontally-related companies developed strong cooperation, but there was only a weak sense of mutual influence (at a weak+ level) and there was less interdependence (at a medium+ level). These findings suggest that cooperation on R&D takes place with less joint development and more independence. In some extreme cases, they conducted separate portions of their R&D agendas independently. But because of their interests in sharing technology and R&D costs and in developing highly advanced technology, CCC interaction in R&D was still based on cooperation. Hence, their interaction seems to have been harmonized under cooperation-oriented CCC governance.

Only 2.2% of the R&D budget was allocated to R&D with vertically-related companies. Due to the smaller operations of vertically-related companies, their contributions to long-term R&D were very limited. These partners developed a strong cooperation and interdependence, interacted frequently, and had the second highest rate of interaction. But semiconductor manufacturers exercised suppressive control vis-á-vis vertically-related companies (a strong+ level of capability to coerce and a weak+ level of vulnerability to be coerced with a score difference of 2.00). This score difference was larger than the difference in any other category in other functional areas. The asymmetric relationship is understandable, since financially and technologically less competent vertically-related companies were dependent on semiconductor manufacturers' help in sharing costs and technology. Their interaction seems to have been harmonized under the rules and norms of hierarchical governance, not under cooperation-oriented CCC interaction.

Only 0.6% of the R&D budget was allocated to companies in the spot market. Company U contracted with companies in the spot market, especially U.S. companies, for semiconductor design and software development, targeting about one and a half years ahead. But the interaction was in many cases based on a one time contract without much possibility for a renewal. These contracts were limited so that semiconductor companies could quickly utilize the software capabilities of other companies. Hence, companies operating in the spot market developed weak cooperation and interdependence, and interacted with semiconductor companies at the least frequent rate. Even so, in R&D interactions, companies in the spot market and semiconductor manufacturers mutually and equally influenced each other to some extent (medium levels of capability to coerce and vulnerability to be coerced). Spot interaction did not show this level of mutual influence in any other functional area. These organizations' interactions clearly showed the characteristics of market governance, except that in this case interacting partners influenced each other to some extent.

Thus, in all R&D areas except spot interaction, partners showed quite high scores in cooperation. Intrafirm R&D in particular indicated very high degrees of interdependence, cooperation, and mutual influence, and these relations were equipped with the capability to maintain technological secrecy. Thus, hierarchical governance may be the most suitable way for harmonizing R&D related actors. Horizontally-related companies did cooperate to maintain their technological leadership, but they tended to undertake R&D more independently. The most vulnerable position, even in comparison with other functional areas, was that of vertically-related companies. Their weak financial and technological capability forced them to be strongly dependent on semiconductor manufacturers when involved in R&D, making the relationship look rather close to hierarchical governance. Companies in the spot market were only for quick acquisitions of minor technology.

C. Summary

Examining the extent to which semiconductor manufacturers transacted with each type of partner in each functional area, I have found that more than seventy percent of transactions were conducted with vertically- and horizontally-related companies for procuring parts, materials and equipment. CCC interaction was predominant. These functional areas involved human relations extensively, which meant that CCC interaction with its emphasis on cooperation, worked very effectively in the context of Japan. But for R&D, most of the work was conducted inside each company, and internalized interaction predominated. Then, what was the nature and characteristics of interaction with each type of partner in each functional area?

Basically, a semiconductor manufacturer and an interacting company, whether small-, medium-, or large-sized, were free to decide how to interact with each other. But the diverse conditions found with each type of partner and in each functional area restricted the range of choices in interactive behavior. Hence, they usually developed certain recognizable characteristics in their interactions. In some cases, the characteristics of interaction were predetermined by diverse existing restrictions, or in some cases interacting partners developed patterns that harmonized their interaction in a set way. Naturally, there were deviant cases. It was even possible for partners to create and modify the environments in which they interacted with each other.

Table 3.6 describes the overall characteristics of interaction. Irrespective of functional area, there was a uniform pattern in developing cooperation and interdependence among different types of partners. The levels of cooperation and interdependence went from weak to very strong in this order: spot, horizontal, vertical, and internalized interactions. In the procurement of parts, materials, and equipment, all types of partners, other than companies in the spot market, could handle asset-specific products, while companies in the spot market simply handled standardized products. In contrast, the asset specificity of products involved in sales activities was weak, and a semiconductor developed for a company did not generate a locking effect in buyer and seller relations. Hence, the semiconductor market was open for newcomers, and competition remained quite severe. No significant pattern was found in the frequency of interaction.

The nature and characteristics of each type of interaction were as follows. Semiconductor manufacturers and vertically- and horizontally-related companies developed a mode of CCC interaction with an intermediate level of cooperation and interdependence. They handled asset-specific products to a medium extent, and mutually influenced each other. Semiconductor companies tended to develop stronger cooperation and interdependence with vertically-related companies than with horizontally-related ones. Semiconductor manufacturers' internalized interaction among intrafirm units was characterized by very strong cooperation and interdependence, and these partners mutually influenced each other. Companies in the spot market were hardly capable of developing cooperative interfirm relationships.

Table 3.6 suggests there are two major factors behind the different patterns that interacting partners develop: the capability to develop cooperation

Table 3.6 Overall Characteristics of Interaction

	TYPE OF INTERACTION			
CHARACTERISTIC	**SPOT INTERACTION**	**HORIZONTAL INTERACTION**	**VERTICAL INTERACTION**	**INTERNALIZED INTERACTION**
General Characteristics in All Functional Areas	weak / weak+ levels of cooperation and interdependence	medium/ medium+ levels of cooperation and interdependence	medium+ / strong+ levels of cooperation and interdependence	strong+ / very strong levels of cooperation and interdependence
Asset Specificity of Products				
Procurement	weak / weak+ levels	medium level	medium level	medium level
Sales	weak	weak	weak	weak
Mutual Influence				
Parts and Materials	no mutual influence	favorable mutual influence for semiconductor companies	favorable mutual influence for companies; highest capability to influence	suppressive control by semiconductor divisions
Manufacturing Equipment	no mutual influence	symmetric mutual influence	suppressive control by semiconductor companies, but frequent interaction	favorable mutual influence for semiconductor divisions
Sales	no mutual influence	unfavorable mutual influence	symmetric mutual influence	favorable mutual influence for semiconductor divisions
R&D	symmetric mutual influence	no mutual influence	suppressive control	favorable mutual influence for semiconductor divisions

and interdependence and the way organizations mutually influence each other. But the ways in which interacting partners influence each other is actually quite complex. Besides, this table does not differentiate between cooperation-oriented and competition-oriented CCC interactions. Table 3.7 combines these three factors and examines patterns.

Intrafirm relations in the areas of R&D and parts, materials, and equipment procurement were well supported by developing very strong cooperation and interdependence. Semiconductor divisions stood in a position that enables them to exercise strong control over intrafirm parts and material supply units. This suppressive control with strong cooperation was common in bureaucratic organizations, when interacting units were minor supporting units such as parts and material bulk purchasing agents. In contrast, semiconductor divisions and intrafirm R&D-related and equipment manufacturing units mutually influenced each other with semiconductor divisions holding a slightly stronger power position. Equipment manufacturing units also engaged in joint R&D for developing highly sophisticated manufacturing equipment for a future-generation of semiconductors. This pattern of mutual influence may have developed because R&D tends to require very strong cooperation and mutual stimulation, while strong bureaucratic control can enable companies to maintain corporate secrets. Individual organizations seem to have provided favorable environments for R&D, and intrafirm actors were harmonized well under hierarchical governance. This judgement is based simply on the factors of cooperation and mutual influence and does not take flexible-synergy and market-like effects into account; therefore, it may be immature. But this interpretation can be also understood as a proposition to be examined in Chapters Four and Five.

Semiconductor manufacturers' interaction with companies in the spot market was found to have the weakest sense of cooperation and interdependence. Spot-market partnerships did not develop any significant level of cooperation or interdependence, and did not involve mutual influence. Sales activities did present an exception to this pattern, however. As explained earlier, the stronger bargaining position of semiconductor companies was the consequence of U.S.-Japan trade disputes, which drastically reduced supply to the gray market handled by semiconductor traders. Other than that, the characteristics of spot market interactions fit well to a general pattern of how market governance harmonizes interaction among market actors.

All other interactions showed CCC with levels of cooperation and interdependence ranging from medium to strong+ levels. In the table, the styles of interaction are divided into three groups. The darkest shaded areas in Table 3.7 show competition-oriented CCC interaction, and light shaded ar-

Table 3.7 Governance of Interaction in Functional Areas

COOPERATION / INTERDEPENDENCE*	SUPPRESSIVE CONTROL SEMICONDUCTOR COMPANIES >> INTERACTING COMPANIES	FAVORABLE MUTUAL INFLUENCE SEMICONDUCTOR COMPANIES > INTERACTING COMPANIES	SYMMETRIC MUTUAL INFLUENCE SEMICONDUCTOR COMPANIES = INTERACTING COMPANIES	UNFAVORABLE MUTUAL INFLUENCE SEMICONDUCTOR COMPANIES < INTERACTING COMPANIES
Very Strong (4.5 – 5.0)	Hierarchical Governance (Internalized Interaction in Parts & Materials)	Hierarchical Governance (Internalized Interaction in R&D)	Hierarchical Governance (Internalized Interaction in Manufacturing Equipment)	
Strong+ (4.0 – 4.4)		Cooperation-oriented CCC Governance (Vertical Interaction in Parts & Materials)	Hierarchical Governance? (Internalized Interaction in Sales)	
Strong (3.5 – 3.9)	Hierarchical Governance? (Vertical Interaction in R&D)	Cooperation-oriented CCC Governance? (Vertical Interaction in Manufacturing Equipment)	Cooperation-oriented CCC Governance (Horizontal Interaction in R&D)	
Medium+ (3.0 – 3.4)		Cooperation-oriented CCC Governance (Horizontal Interaction in Parts & Materials)	Competition-oriented CCC Governance (Vertical Interaction in Sales)	Competition-oriented Governance (Horizontal Interaction in Sales)
Medium (2.5 – 2.9)			Cooperation-oriented CCC Governance? (Horizontal Interaction in Manufacturing Equipment)	
Weak+ (2.0 – 2.4)			Market Governance (Spot Interaction in Parts and Materials)	
Weak (1.5 – 1.9)	Market Governance (Spot Interaction in Sales)		Market Governance (Spot Interaction in R&D) Market Governance (Spot Interaction in Manufacturing Equipment)	

Note: CCC: Competitive-*cum*-cooperative.   * An averaged score of Degrees of Cooperation and Interdependence was used to indicate cooperation/interdependence.

eas indicate cooperation-oriented CCC interaction. The remaining types of CCC interaction are indicated with question marks, suggesting that they may be deviant cases or evidence is insufficient.

Semiconductor companies in sales activities developed competition-oriented CCC interaction with horizontally- and vertically-related customers. In an environment of severe market competition, which was characterized by the weak asset specificity of products, semiconductor manufacturers tried to develop cooperation with their customers in order to obtain crucial information on new products, suggest their ideas for future developments, and win contracts. However the highest level of cooperation they could achieve was a medium+ level. Cooperation beyond this level may not have been beneficial to these customers, since to develop stronger cooperation requires to reduce the number of interacting companies and weakens the capability of a company to take advantages of seller-initiated competition among semiconductor suppliers over cooperative behavior. Semiconductor companies stood in a weaker bargaining position vis-á-vis these customers. Horizontally-related customers in particular, being highly attractive to semiconductor companies because of their ability to make bulk purchases, to develop new technology and products, and to ensure a longer-duration of product sales, had a strong bargaining position. But their interaction did generate mutual influence. Even vertically-related companies, which were usually vulnerable to the influence of semiconductor manufacturers in other functional areas, stood in an equal bargaining position as customers, and established a mutual influence. The evidence from questionnaires and interviews seems to indicate the severity of the market competition and the fact that semiconductor manufacturers used cooperation as a means of reducing uncertainty and competition in the market. Since interaction between semiconductor companies and these customers delicately mixed cooperation and competition, I conclude that their interaction was harmonized under CCC governance. But the relationship was competition-oriented, because cooperation was used as a means of reducing uncertainty created by severe market competition. Chapter Five provides further evidence to prove this point.

The light-shaded areas in Table 3.7 indicate partnerships with the typical characteristics of cooperation-oriented CCC interaction. Vertically- and horizontally-related companies in parts and materials procurement developed an intermediate level of cooperation and interdependence with semiconductor companies, handled an intermediate level of asset-specific products, and worked under conditions of mutual influence with semiconductor companies, in which the semiconductor companies held a slightly stronger power position. Given the importance of quality, cost, and delivery control, cooperation became of primary importance for semiconductor companies. Hence, they engaged in cooperation-oriented CCC interaction. Similarly, horizontally-related companies developed strong cooperation and interdependence and engaged in joint R&D. Because of their highly sophisticated technological capabilities, they could stand in an equal bargaining position vis-á-vis semiconductor manufacturers, and develop conditions of mutual influence. Since good cooperation is a key to successful joint R&D, CCC interaction became cooperation-oriented.

The cases of vertical interaction in R&D and manufacturing equipment procurement were unusual. Semiconductor manufacturers strongly controlled vertically-related companies, while the latter were highly cooperative and dependent on the former in R&D activities as well as in the sales of their manufacturing equipment. The suppressive control and strong cooperation and interdependence may be characterized as a part of a bureaucratic system, rather than as CCC interaction, since strong cooperation prevents the effective implementation of competition-generating measures. Chapters Four and Five examine this idea further.

A puzzle also remains regarding intrafirm sales activities. Semiconductor divisions and intrafirm customers developed very strong interdependence, but their sense of cooperation was slightly lower than the level of interdependence. This discrepancy was found only in this functional area. These intrafirm partners also mutually influenced each other, with semiconductor divisions holding a slightly stronger power position. Their business in high-risk IC development may explain their highly intensive interaction and the favorable mutual influence. However, the weak level of asset specificity may suggest that their products were not different from those handled by vertically- and horizontally-related companies. Both semiconductor divisions and intrafirm customers operated as profit centers, and they interacted like independent companies. This type of operation created a market-like environment inside a company, and allowed the divisions to operate outside of the usual bureaucratic ways of allocating resources. In Table 3.7, a question mark is attached to internalized interaction in sales, suggesting the need for further inquiry in Chapter Four.

Horizontally-related equipment manufacturers were in long-term relationships with semiconductor companies. These partners were mainly linked by technological needs. Although these equipment manufacturers had the most frequent levels of interaction with semiconductor companies and both partners equally influenced each other, their areas of cooperation were limited, and the level of cooperation was even lower than that found in competition-oriented CCC interaction. This finding raises the question of whether horizontally-related companies developed true cooperation-oriented interaction with semiconductor companies. The question mark in the table suggests the need for further inquiry in Chapter Four.

These findings suggest that CCC interaction has diverse characteristics. While the cooperation-oriented CCC interaction in the procurement of parts and materials and the competition-oriented CCC interaction in sales are the model cases for this research, the interactions in functional areas other than R&D, in general, seem to show CCC characteristics. But deviant cases were also found. Some CCC interaction seems to have developed the characteristics of internalized interaction, while some internalized interaction may have operated in ways similar to cooperation-oriented CCC interaction. Evidence in the next two chapters confirms my interpretations and clarifies many of the puzzling questions raised in this chapter.

Chapter Four

Effects of Cooperation-oriented CCC Interaction in Procurement Relations

In Chapter Three, I demonstrated that competitive-cum-cooperative (CCC) interaction was dominant in the procurement of parts and materials (81.0%) and manufacturing equipment (73.9%). CCC interaction has some unique characteristics that fit well with most modes of interaction but were not well nurtured in spot or internalized interaction. The power of CCC interaction lies in the way it delicately blends flexible-synergy and market-like effects.

Cooperation-oriented CCC interaction is characterized by medium to strong+ levels of cooperation, significant mutual influence, and intermediate levels of asset-specific product relations. These characteristics are outlined in Chapter Three. Because it relies on flexible cooperation, CCC interaction generates flexible-synergy effects. But partners working at this level of cooperation also accept competition-generating measures, a contradictory element. Competition-generating measures, implemented to an extent that does not hamper cooperation, enhances sensitivity to market movements and flexibility so that companies can cope with changes, generating market-like effects. A combination of the two approaches gives comparative advantages that cannot be generated by other types of interaction.

The capability to generate flexible-synergy and market-like effects would differ by type of interacting partner. Interaction between vertically-related suppliers and semiconductor manufacturers would tend to reach higher levels of flexible cooperation and flexible-synergy effects and fewer market-like effects. In contrast, owing to an emphasis on performance, interaction between horizontally-related suppliers and semiconductor manufacturers would generate more market-like effects. This pattern is examined in this chapter.

Managers in the various functional areas of semiconductor firms were asked to identify the most important company that they interacted with in spot, internal, vertical, and horizontal relations. The managers used several measures to assess the benefits their companies gained from long-term mutual accumulation, satisfying mutual expectations, and future-oriented risk-taking activities.[1] These dimensions constituted a portion of their evaluation criteria based on the logic of continuity.

To examine the extent of market-like effects, I compared the degrees of benefits that semiconductor managers reported gaining from implementing competition-generating measures vis-á-vis each type of partner.[2] Managers assessed the degree to which their companies could benefit from adjusting their inter- or intrafirm relations according to fluctuation in the market, by such measures as implementing strategic pricing, creating a competitive environment through strategies like introducing other suppliers and negotiating to purchase inexpensively (benefits from market-linked adjustments). Managers also reported on the degree to which their companies could flexibly adjust resource interdependence according to market movements, with

---

[1] Respondents were asked to select the most important company representative for each type of interaction, and to answer the following questions:

(1) What is the degree of benefit that your company enjoys from the indicated company in the following issues?
(a) Acquiring technological information
    (1. hardly any, 2. a little, 3. some, 4. a good deal, 5. a great deal)
(b) Acquiring product-market information
    (1. hardly any, 2. a little, 3. some, 4. a good deal, 5. a great deal)
(c) Obtaining business, owing to long-term mutual experience
    (1. hardly any, 2. a little, 3. some, 4. a good deal, 5. a great deal)
(d) Enhancing trust and cooperation with each other
    (1. hardly any, 2. a little, 3. some, 4. a good deal, 5. a great deal)
(e) Coordinating delivery and quality control
    (1. hardly any, 2. a little, 3. some, 4. a good deal, 5. a great deal)
(f) Receiving stable supplies
    (1. hardly any, 2. a little, 3. some, 4. a good deal, 5. a great deal)
(g) Receiving services that satisfy your company's needs
    (1. hardly any, 2. a little, 3. some, 4. a good deal, 5. a great deal)
(h) Keeping corporate secrets
    (1. hardly any, 2. a little, 3. some, 4. a good deal, 5. a great deal)
(i) Stimulating creativity
    (1. hardly any, 2. a little, 3. some, 4. a good deal, 5. a great deal)
(j) Developing future-oriented risk-taking projects
    (1. hardly any, 2. a little, 3. some, 4. a good deal, 5. a great deal)
(k) Coordinating the development of new manufacturing equipment
    (1. hardly any, 2. a little, 3. some, 4. a good deal, 5. a great deal)
    [Note: This question was only asked on a questionnaire concerning the procurement of manufacturing equipment.]
(l) Coordinating the maintenance and services of purchased manufacturing equipment
    (1. hardly any, 2. a little, 3. some, 4. a good deal, 5. a great deal)
    [Note: This question was only asked on a questionnaire concerning the procurement of manufacturing equipment.]

the flexible application of contracts used in just-in-time production (benefits from resource interdependence adjustments). In addition, they evaluated the degree to which strict standards and tough relations were maintained (benefits from reevaluation adjustments). This is an important measure because effective evaluation allows semiconductor companies to achieve diverse benefits.

Questions for flexible-synergy effects were designed to cover the diverse conditions of intra- and interfirm cooperation, while those for market-like effects covered the diverse conditions of market and market-like environments. In other words, questions were written to cover the diverse possibilities of interactive behavior. Each questionnaire was answered by a different manager related to each functional area. In analyzing the data, I used the same criteria used in Chapter Three.

A.   Procurement of Parts and Materials

1.   Flexible-synergy Effects

a.   Benefits from Long-term Mutual Accumulation

Through long-term relations, cooperating partners accumulate know-how and develop effective means of exchanging information. This kind of long-term mutual accumulation provides an easier environment for partners to conduct more business together, resulting in further enhancing cooperation and trust.

---

[2] Respondents were asked to select the most important company representative for each type of interaction, and to answer the following questions:

(1) What do you perceive is the degree of benefit that your company enjoys from the indicated company on the following issues?
  (a) Implementing strategic pricing (planned reduction of prices)
     (1. hardly any, 2. a little, 3. some, 4. a good deal, 5. a great deal)
  (b) Creating a competitive environment by introducing another company that deals with similar products
     (1. hardly any, 2. a little, 3. some, 4. a good deal, 5. a great deal)
  (c) Purchasing inexpensively
     (1. hardly any, 2. a little, 3. some, 4. a good deal, 5. a great deal)
  (d) Altering procurement sources flexibly
     (1. hardly any, 2. a little, 3. some, 4. a good deal, 5. a great deal)
  (e) Implementing contracts flexibly, according to market conditions
     (1. hardly any, 2. a little, 3. some, 4. a good deal, 5. a great deal)
  (f) Maintaining strict standards and a tough relationship
     (1. hardly any, 2. a little, 3. some, 4. a good deal, 5. a great deal)
  (g) What are some of the methods you used to evaluate the indicated company? (Please list concretely.)
     (   )

Table 4.1 Flexible-synergy Effects by Type of Interaction in Procurement of Parts and Materials

BENEFITS	TYPE OF INTERACTION			
	SPOT INTERACTION (N = 6)	HORIZONTAL INTERACTION (N = 5)	VERTICAL INTERACTION (N = 6)	INTERNALIZED INTERACTION (N = 6)
A. From Long-term Mutual Accumulation				
Acquiring Technological Information*	2.83	3.40	3.67	3.00
Acquiring Product-market Information*	2.83	3.00	3.17	2.80
Obtaining Business, Owing to Long-term Mutual Experience*	2.67	3.40	3.67	3.25
Enhancing Trust and Cooperation*	2.83	3.40	3.50	3.80
B. From Satisfying Mutual Expectations				
Coordinating Delivery and Quality Control*	2.67	3.60	3.83	3.80
Receiving Stable Supplies*	3.00	3.60	3.83	3.20
Receiving Services to Satisfy Needs*	3.00	3.20	3.67	3.00
Keeping Corporate Secrets*	2.50	3.60	3.33	4.40
C. From Future-oriented Risk-taking Activities				
Stimulating Creativity*	2.67	3.20	3.50	2.80
Developing Future-oriented Risk-taking Projects*	2.67	3.40	3.83	3.20
D. Common Goals and Interests				
	-Mutual profit -Cost	-Mutual profit -Cost -Mutual prosperity -Development of new technology -Information -Technology	-Mutual profit -Cost -Mutual prosperity -Development of new technology -Information -Technology	-Mutual profit -Mutual prosperity

Note: *1.0 means very weak and 5.0 means very strong. Scores are categorized as follows: very weak (mean value = 1.0 - 1.4); weak (1.5 - 1.9); weak+ (2.0 - 2.4); medium (2.5 - 2.9); medium+ (3.0 - 3.4); strong (3.5 - 3.9); strong+ (4.0 - 4.4); and very strong (4.5 - 5.0).

To assess the benefits gained from long-term mutual accumulation, four questions were asked; the questions asked managers to rate how much they gained in the areas of technology and product-market information acquisition, additional business owing to long-term relations, and enhanced sense of trust and cooperation.

According to Table 4.1, vertically-related suppliers made especially good use of long-term relations for promoting business (at a strong level), while intrafirm parts/material supply units and horizontally-related suppliers did so a little less (at a medium+ level). This benefit is extremely important for semiconductor companies, because continued and expanded business relations are a vital means to achieve their shared goals and interests in mutual prosperity with these three types of partners (see Common Goals and Interests in Table 4.1). Suppliers in the spot market could make only a little use of long-term relationships (at a level of medium), and did not share the goal of mutual prosperity.

One benefit of smoother long-term relations is the exchange of technological information. Vertically- and horizontally-related suppliers were more effective in providing technology information to semiconductor manufacturers than any other type of partner (at medium+ to strong levels). Vertically-related suppliers were especially effective. As the list of common goals and interests in Table 4.1 shows, acquiring technology information was considered one of the major reasons why semiconductor companies cooperated with vertically- and horizontally-related suppliers. Intrafirm relations and suppliers in the spot market were less capable of providing such benefits. However, no clear difference existed between partners in providing market information to semiconductor companies.

Managers in Companies R, Q and T strongly emphasized the importance of CCC interaction in exchanging technological information. Unless a company has a good flow of technological information from parts and material suppliers, it will miss highly specialized valuable technological information, stimulation for new ideas, and market opportunities (Toyo Keizai, 1982:20-21). The computer-integrated operations between suppliers and semiconductor manufacturers, mainly undertaken for problem solving, actually function to ensure companies exchange the most relevant and suitable information. Managers even reported having periodic technology information exchange meetings.

Companies were after any type of fragmented information that could give hints about the types of integrated circuits that were being developed by competitors, how other companies were reducing cost and failure rates, and how competitors were improving quality (Aida, 1992:372). For example, a statement such as "Company G increased its purchase of a certain material" can suggest product-market movements. The accumulation of fragmented information like this, in addition to a company's own information, can form the basis for interpreting product-market movements. Thus, companies require extensive interpersonal networks, of which vertically- and horizontally-related suppliers are the core.

Does this mean, then, that CCC interaction necessarily enhances trust and cooperation? CCC interaction (at medium+ to strong levels) enhanced trust and cooperation, prevented opportunism, and reduced uncertainty more than spot interaction in a market (at a medium level), but less than intrafirm relations (at a strong level). A slightly weaker level of trust and cooperation may indicate the strength of CCC interaction, since it can make the implementation of competition-generating measures possible.

b.   *Benefits from Satisfying Mutual Expectations*

One of the toughest criteria for maintaining a long-term relationship is each partner's success in satisfying mutual expectations. The level of achievement required is usually extremely high, since, to keep winning in competition, improvements in areas including production, quality, cost, and delivery have to be implemented beyond competitors' achievements (Gregory, 1985; Sakamoto, 1990:18-23). Four questions were asked to measure the benefits of satisfying mutual expectations. These questions addressed the benefits of receiving stable supplies, coordinating delivery and quality control, receiving services to satisfy semiconductor companies' needs, and keeping corporate secrets.

Both horizontally- and vertically-related suppliers, according to Table 4.1, showed a strong tendency to keep supply relations stable (at a strong level), while purchases from intrafirm supply units and suppliers in the spot market tended to involve a little less stability in supply (at a medium+ level). Vertically-related suppliers were especially strong in providing more services that satisfy semiconductor companies' needs, while horizontally-related suppliers did less so, but still did slightly more than intrafirm supply units (both at a medium+ level). These findings suggest that vertically- and horizontally-related suppliers, especially the former, were effective in satisfying mutual expectations in both supply stability and services. Though it is not a particular characteristic of CCC interaction, vertically-related suppliers, along with intrafirm supply units, strongly coordinated delivery and quality control. Horizontally-related companies coordinated these activities a little less effectively (all of the three partners at a strong level).

Vertically-related suppliers satisfied mutual expectations a little better than other types of partners did. Several factors compelled semiconductor manufacturers to give priority in purchasing to vertically-related suppliers. First, vertically-related suppliers responded favorably to the difficult and urgent needs of semiconductor manufacturers. Since silicon cycles often put companies into turbulent conditions, sometimes causing oversupply and sometimes causing shortages, vertically-related suppliers' cooperation became very valuable (Nikkei Business, 1984a:27). Second, constant improvements were an important part of satisfying mutual expectations and were sometimes achieved without involving monetary exchanges. For example, a small segment of design is often changed without charging extra money under the name of services. Practice like this actually made tiny improvements much easier to achieve than in cases in which every change required negotiations and charges.[3] There is clear evidence that vertically-related suppliers were the source of extra benefits for satisfying mutual expectations.

The favorable behavior that vertically-related suppliers offered to semiconductor companies cannot be explained simply as exploitation made possible by power asymmetry. To assure such benefits, semiconductor companies needed to transfer technology and provide financial help, training, and managerial advise, in addition to other help. Such costs cannot be ignored. Though power asymmetry enabled semiconductor manufacturers to create favorable conditions, I argue that interaction was primarily based on cooperative bargaining between the two partners. Vertically-related suppliers enjoyed the benefits of stable purchases and diverse help, while semiconductor manufacturers benefited from flexibility in production and extra services to satisfy needs quickly.

The best evidence to support my argument comes from times of crisis. When a vertically-related supplier becomes technologically competent, a semiconductor manufacturer comes to depend on it. But if the supplier has to lay off some competent workers at a time of drastic decline in demand, the semiconductor manufacturer will even make extra purchases at a higher price to keep the supplier alive.[4] For the semiconductor manufacturer, lay-offs can actually mean the supplier's loss of technology and reliability, resources for which they have usually invested enormous amounts of time, energy, and money. For example, when the semiconductor market faced a surplus, prices rapidly went down, and production had to be cut. Given a decline in demand, Company Q could have requested a uniform production cut of all vertically-related suppliers. Instead, it designated a proportion of production to be cut, according to the importance of their technology and with an eye to potential risks that the production cut could cause.

The managers in this study considered maintaining technological secrecy between companies quite important, especially because the semiconductor industry deals with highly sophisticated technology as well as easily transferable ideas.[5] Intrafirm supply units were the most successful at tightly controlling technology secrecy. Their score was the highest at a strong+

---

[3] Interview with Company T manager.
[4] Interview with Company Q manager.
[5] Interview with Company T manager.

level. This is perhaps why the most sophisticated technology was in many cases developed within each company. Technologically competent horizontally-related suppliers also maintained technology secrecy well (at a strong level), showing a slightly higher score than vertically-related suppliers (at a medium+ level).

Regardless of who the business partner was, secrecy over highly sophisticated technology was contractually controlled. Otherwise, companies would lose the basis for a trusting relationship. Some horizontally-related suppliers, operating in oligopolistic parts and material markets, catered to multiple semiconductor companies and needed to establish credibility with each manufacturer. For example, a photo mask could reveal the top secrets of products. Photo mask manufacturers in this situation arranged their organizational structures so that a fixed group of workers would work for only one customer. Workers in the group were given hardly any opportunity to chat or even to meet other group members who dealt with different customers.[6] In this sense, vertically-related suppliers' relatively weak control over technology secrets may have been due to the less sophisticated nature of the technology that they handle.

c.   *Benefits from Future-oriented Risk-taking Activities*

The most difficult criteria to meet were the stimulation of creativity between interactive partners, the serious development of future-oriented risk-taking projects, and success in developing new technology or products. In order to maintain long-term relationships, companies had to cooperate from an early phase in developing future technology and products with their own risk. Two questions on the survey identified benefits derived from future-oriented risk-taking activities by asking whether partners stimulated creativity and whether they actually developed projects.

According to Table 4.1, partners in CCC interaction stimulated creativity and developed future-oriented risk-taking projects better than any other types of partners. Vertically-related suppliers brought especially strong benefits in both areas, while horizontally-related ones did so to a slightly lesser extent (both at a medium+ level). This was a highly important aspect of flexible-synergy effects, since semiconductor companies identified the development of new technology as one of the shared goals and interests with both vertically- and horizontally-related suppliers (see Common Goals and Interests in Table 4.1). Internalized interaction was effective to some extent in devel-

---

[6] Interview with Company Q manager.

oping future-oriented risk-taking projects (a medium+ level), but weak in stimulating ideas (a medium level). Their weak ability to stimulate innovation was because intrafirm supply units are, in general, minor supporting units, not much involved in future-oriented activities, unless they produce some specialized parts and materials. In contrast, spot interaction received the lowest scores in both categories (at a medium level).

Interfirm relations often became quite helpful, since the internal development of supply capability often became quite costly and risky due to the involvement of highly specialized technology. One example is SONY's case. SONY's success with the silicon transistor was greatly owing to the development of a CCC relationship with Mitsubishi Materials Silicon (formerly Japan Silicon). When the president of Mitsubishi Materials Silicon visited SONY to discuss the development of technology dealing with germanium, the president of SONY recommended him to develop silicon, predicting its future importance. SONY gave this important advice, because it did not have a chemical background and the development was too risky for SONY to engage in. But on top of these calculated reasons, SONY trusted that the company would do its best to comply with the demand, because of their long-term relationship. Although production demanded a high degree of purity for the silicon crystal it used, and the cost of its production became quite high, the company complied with the demand (Nakagawa, 1985:124). As well, SONY committed to continue purchasing the silicon crystal from this company.

One caution, however, is necessary. CCC future-oriented risk-taking activities do not necessarily mean a contract, which is made only at the mass production stage. There is no guarantee that future-oriented activities will result in a long-term relationship. Only companies successful in developing high quality products and responding to semiconductor manufacturers' needs are rewarded with a contract. But the probability of winning one becomes very high, since only a few companies are given opportunities to engage in joint activities.

Thus, in the procurement of parts and materials, CCC interaction generated better flexible-synergy effects than any other type of interaction. It showed higher benefits in both of the indicators of future-oriented risk-taking activities. This criteria seems to be quite important, for developing new technology was one of the stated goals shared with vertically- and horizontally-related suppliers. The importance of satisfying mutual expectations and long-term accumulation differed according to the issue and the type of partner. Both vertically- and horizontally-related suppliers seemed to respond better to providing stable supplies and services to satisfy needs, which were considered to be the extra needs of semiconductor companies. Intrafirm supply units responded poorly to fluctuation in needs. Having the mutual expectation that a partner will satisfy an unexpected need was an important characteristic of CCC interaction. Also, as a shared goal, long-term accumulation allowed semiconductor companies to use vertically- and horizontally-related suppliers as the core of their technology information networks. Vertically-related suppliers benefited the most by turning long-term relations into business opportunities. Though not as conspicuous as the other two characteristics of the logic of continuity, benefits derived from long-term mutual accumulation were also an important part of CCC interaction.

As a whole, five indicators out of nine suggested the strength of both vertically- and horizontally-related suppliers. In contrast, internalized interaction generated the strongest benefits in only one indicator of long-term mutual accumulation and one indicator of satisfying mutual expectation. Spot interaction did not generate the best benefits in any category.

2.   Market-like Effects

a.   Benefits from Market-linked Adjustments

Semiconductor companies developed diverse measures to adjust interfirm relations to market movements. Three well-known measures are strategic pricing, multiple sourcing, and altering procurement sources, sometimes temporarily or sometimes permanently. In strategic pricing, interacting companies cooperatively developed a schedule for reducing prices. In multiple sourcing, semiconductor companies purchased the same parts and materials from a few suppliers, so that competition took place among them. The effectiveness of such measures was tested by asking managers to indicate the degree of benefits gained from implementing strategic pricing, creating a competitive environment by introducing another supplier, purchasing inexpensively, and altering procurement sources flexibly.

Purchasing parts and materials inexpensively was a very attractive benefit for semiconductor manufacturers. According to Table 4.2, vertically-related suppliers were the best partners for purchasing parts and materials inexpensively (at a level of strong). Horizontally-related suppliers also showed a score quite close to vertically-related suppliers (at a medium+ level) and higher than intrafirm supply units.

This finding raises an interesting question. According to neo-classical economics, market governance is operated under the principle of competition among self-interested companies. It is supposed to be most effective at creating competition and pressuring companies to lower prices. Contrary to

Table 4.2 Market-like Effects by Type of Interaction in Procurement of Parts and Materials

BENEFITS	TYPE OF INTERACTION			
	SPOT INTERACTION (N = 5)	HORIZONTAL INTERACTION (N = 5)	VERTICAL INTERACTION (N = 5)	INTERNALIZED INTERACTION (N = 5)
A. From Market-linked Adjustments				
Implementing Strategic Pricing*	2.67	3.40	3.00	3.00
Creating Competitive Environment by Introducing Another Company*	3.17	3.20	3.17	2.80
Purchasing Inexpensively*	3.17	3.40	3.50	3.00
Altering Procurement Sources Flexibly*	3.33	3.00	2.67	2.40
B. From Resource Interdependence Adjustments				
Implementing Contracts Flexibly*	2.67	3.00	3.00	3.20
C. From Reevaluation Adjustments				
Maintaining Strict Standards and Tough Relationship*	2.50	3.40	3.20	3.40

Note: * 1.0 means very weak, and 5.0 means very strong. Scores are categorized as follows: very weak (mean value = 1.0 - 1.4); weak (1.5 - 1.9); weak+ (2.0 - 2.4); medium (2.5 - 2.9); medium+ (3.0 - 3.4); strong (3.5 - 3.9); strong+ (4.0 – 4.4); and very strong (4.5 - 5.0).

this theory, this finding suggests that CCC interaction can generate a lower price than the spot market. It is important to note that parts and materials markets are generally less competitive than the semiconductor market, and that the need for tight control over quality requires cooperation between buyers and sellers. According to neo-classic economics, an oligopolistic market with colluded practices raises prices. Then prices in these transactions should be higher than those found in the spot market. Even if we assume that the spot market and CCC interaction have similar levels of competition, then the price should not be different. But the fact that CCC interaction can generate a lower price means that there is some mechanism at work that creates severer competition than the market without losing the benefit of cooperation. In other words, CCC interaction serves two functions: it maintains cooperation, which is crucial for controlling the quality of parts and materials, and it pressures prices lower than the spot market. This finding clearly indicates the power of CCC interaction.

Strategic pricing was another competition-generating measure that kept parts and materials prices lower. Strategic pricing relied on estimations of the effects of a learning curve to make a price reduction plan for the various stages of production. One way to do this was to calculate a payment by simply multiplying a number of processed pieces by the sum of estimated costs of every process. When this formula was used, parts and materials suppliers often made every effort to squeeze profits by reducing the costs of every process. They employed more part-timers, subcontracted the production of more parts, reduced the actual number of processes, and automated production procedures. Another way they lower costs was to use more detailed information from their partners, including a mutually-agreed-upon range of profits and costs for administration and personnel (Asanuma, 1985). One example of the effects of strategic pricing was a company's redesign of a corner of semiconductor from square to round which was done to reduce the amount of resin usage. The material in some parts of the semiconductor was also changed from gold to gold plate. Detailed information offered by parts and materials suppliers helped the semiconductor manufacturer to make such decisions.[7]

To implement policies like this one, semiconductor manufacturers needed to solicit strong cooperation from partners, request detailed information, have extensive knowledge of the production of parts and materials, and constantly evaluate their partners' capability (Asanuma, 1985). Strategic pricing functioned as an alternative method of generating a market-like environment, but it tended to work more effectively when flexible cooperation was well established.

Horizontally-related suppliers responded best to strategic pricing measures (at a medium+ level); in this case, strategic pricing kept parts and materials prices lower. Vertically-related suppliers and intrafirm supply units also responded to this measure, though not as well as horizontally-related suppliers did (both at a medium+ level). Those in the spot market responded least effectively (at a medium level).

One interesting finding is that while semiconductor manufacturers could purchase parts and materials inexpensively from vertically-related suppliers with a less stringent application of strategic pricing than horizontally-related suppliers required, the latter's lower prices were the consequence of the stringent application of strategic pricing. This was largely because the relationship with vertically-related suppliers was based more on human relations with a strong sense of cooperation and trust. In contrast, the relationship with horizontally-related suppliers involved rigorous negotiation over strategic pricing with a slightly weaker sense of trust and cooperation. Whether the logic of continuity was human-relations or performance-oriented, semiconductor manufacturers could purchase parts and materials more inexpen-

---

[7] Interview with Company Q manager.

sively from vertically- and horizontally-related suppliers than from intrafirm supply units or suppliers in the spot market.

Multiple sourcing, having only a few companies supplying the same parts and materials, introduced competition among the suppliers and created a market-like environment. But by developing cooperative relations with all of the selected companies, this mode of interaction also provided a safety net to ensure a stable supply.[8] All types of partners except intrafirm supply units responded well to the competitive environment that was created by introducing another supplier (a medium+ level). Intrafirm procurement transactions seemed the least responsive to multiple sourcing due to their very strong sense of cooperation. Although market governance is considered the most effective institutional arrangement for generating a competitive environment, these findings indicate that CCC governance can do so as easily as market governance, without losing the benefits derived from cooperative relations. Hence, CCC governance is quite an effective institutional arrangement for generating a market-like environment, while it also provides the benefits of flexible-synergy effects.

In many cases, competition was introduced either to horizontally-related suppliers or to vertically-related ones. Any parts and materials manufacturer failing to comply with standards or demands faced warning or punishment, of temporarily losing some, not all, of its contracts to competitors; the contract would only be returned if the standards of the semiconductor company were satisfied. This method of warning by temporarily canceling a part of a contract was unique to long-term relationships, since the real purpose of this punishment was to warn a partner and provide opportunities for a better long-term relationship. Failure to comply to demands after a certain period of warning clearly meant a termination of the relationship (Howard, 1990).

Sometimes semiconductor companies made vertically- and horizontally-related suppliers compete with each other using the same stringent standards; this strategy was used especially when the drastic improvement of a vertically-related company was needed. For example, a long-term vertical supplier, Company H, lost its contracts with a semiconductor manufacturer, Company Q, for six months. Company H lost the contract because it used dirt dug from shallow earth to produce silicon and failed to remove alpha particles, which created errors when electric current passed through the integrated circuit (IC). Immediately after identifying the cause of their IC failure, Company Q terminated an order and shifted it to a horizontally-related

---

[8] Interview with Company V manager.

company, Company I. Up to this point, Company Q behaved according to the logic of continuity based on performance. However, when Company H responded by building a new factory and using dirt from deeper earth, Company Q reinstated its contracts.

Company Q shifted its order back to Company H for the following reasons: (1) Company H had already developed a strong sense of cooperation with Company Q; (2) they benefited from flexible-synergy effects generated by the accumulation of experience; (3) vertically-related suppliers were much more responsive to Company Q's diverse and tough demands; and (4) Company Q could implement diverse competitive measures more easily with Company H than Company I. This suggests that once performance criteria were met, the logic of continuity based on human relations was weighed more, making the vertically-related supplier the more favorable partner. However, Company I received a different contract, so that Company Q could maintain cooperative relations with the company.

Competition created from multiple-sourcing forced parts and materials suppliers to be more independent of semiconductor manufacturers, though their cooperative relations remained firm. The suppliers also searched for new customers to stabilize their sales. Their business experiences with a different semiconductor manufacturer brought new insights and improvements in the diverse dimensions of their operations, especially in terms of technology. These changes also brought benefits to the old long-term partner.

No other type of partners, however, could match the capability of companies in the spot market to switch procurement sources flexibly. For CCC interaction, terminating a long-term partnership permanently was not a rare incident. For example, in the 1970s Company Q asked a large-sized glass company, Company O, to refine the quality of sheet glass by increasing the accuracy of flatness. But Company O did not respond to this demand, because at that time semiconductor-related products earned a small profit relative to automobile glass. This incident, which resulted in the termination of a long-term relationship with Company O, provided a lucky opportunity for a smaller-sized company. Company Q was also involved in similar incident with companies producing oxygen.[9]

Even well-nurtured cooperative relationships sometimes cannot survive when product-market competition becomes extremely intense. For example, in the 1970s cost competition for calculators became so severe that competi-

---

[9] Interview with Company Q manager.

tors had to resort to automation and/or relocating factories to rural areas. Company F shifted its factory from Nara to Mie Prefecture so that it could terminate costly subcontractors and redevelop less expensive small-sized subcontractors in a remote rural area (Aida, 1992:365). This meant that severe product-market competition can even terminate long-term relationships, though rebuilding a relationship with new partners can be quite costly.

b.   Benefits from Resource Interdependence Adjustments and Reevaluation Adjustments

Market-like effects generated by resource interdependence adjustments were measured by the amount a company benefited from exercising contractual assurance with built-in flexibility, in short, flexible contracting. This practice was developed as a part of just-in-time production. To be more specific, it means that despite the date and amount of delivery specified in a contract, a buyer could request a later delivery or a reduced amount when demand was slacking or early delivery when demand was increasing. Further adjustments could be made with the signing of a new contract. Through this method, a semiconductor manufacturer could adjust the amount of supply delivery according to product-market movements, while it served the channel of supply through long-term relations. Intrafirm supply units responded to the flexible application of contracts the best (at a level of medium+), while vertically- and horizontally-related suppliers scored only a little lower (all of these types of partners were at a medium+ level).

Strict and constant evaluation of long-term suppliers' performance allowed semiconductor companies to propose performance improvements and adjustments, and to enjoy benefits from reevaluation adjustments. The more detailed the specification was, the stricter semiconductor companies tended to make their tests (Ikeda, 1990). Both horizontally-related suppliers and intrafirm supply units responded to the strict standards and tough demands of semiconductor companies (all three types of partners were at a medium+ level). Vertically-related suppliers responded a little less effectively than these two types of partners. This is not necessarily a unique characteristic of CCC interaction, since intrafirm evaluation can also be tough and stringent.

Evaluations were made daily. When a problem occurred, an immediate solution and improvement was expected. If improvement could not be generated within a few days, the semiconductor company would give a yellow light to long-term relations.[10] But semiconductor manufacturers also imple-

---

[10] Interview with Company P manager.

mented additional checks on a partner's willingness to cooperate in many different areas, including production and cutting costs.

Thus, semiconductor manufacturers used diverse competition-generating measures to create market-like environments for cooperating partners. Competition-generating measures unique to CCC interaction were found in interactions that used market-linked adjustments. Both vertically- and horizontally-related suppliers sold their products with low prices, while horizontally-related suppliers also responded well to strategic pricing. Resource interdependence adjustment was more effectively implemented within single company, while reevaluation adjustments were well practiced by both horizontally-related suppliers and intrafirm supply units. These adjustments are not particular characteristics of CCC interaction.

3.   Summary: Dynamics of Cooperation-oriented CCC Interaction

Table 4.3 summarizes performances by type of interaction. The row labeled top performance lists the areas in which each type of interaction scored the highest; the low-performance row lists factors that were rated in the bottom two positions. When the difference between top scores in any one indicator was less than 0.2, then that indicator was listed under both types of partners. When the top three scores were similar, then they were treated as an intermediate level of performance because the similarity in scores indicated a general characteristic of the industry rather than a quality of specific type of partner. When the second and third ranked scores were similar, then both were ranked second. Indicators of an intermediate level of performance are not listed. Regular letters represent indicators used for testing flexible-synergy effects, while italic letters indicate those used for testing market-like effects.

Table 4.3 clearly shows that vertically- and horizontally-related suppliers performed far better than intrafirm supply units and suppliers in the spot market in generating both flexible-synergy and market-like effects. In the low performance category, suppliers in the spot market performed less effectively in nine out of eleven flexible-synergy-effects-related indicators and four out of six market-like-effects-related ones. Intrafirm supply units showed five of the former and three of the latter.

Since vertically-related suppliers had low performances in one indicator of flexible-synergy effects and two indicators of market-like effects, it is more proper to say that horizontally-related companies mixed these two effects better than vertically-related companies, having only one low perform-

Table 4.3 Performance by Type of Interaction in Procurement of Parts and Materials

PERFORMANCE	SPOT INTERACTION	HORIZONTAL INTERACTION	VERTICAL INTERACTION	INTERNALIZED INTERACTION
		TYPE OF INTERACTION		
Top Performance (Top in Ranking)	*Altering procurement sources flexibly*	*Implementing strategic pricing* Purchasing inexpensively* *Maintaining strict standards and tough relationship'*	Acquiring technological information Obtaining business, owing to long-term mutual experience Coordinating delivery and quality control'* Receiving stable supplies Receiving services to satisfy needs Stimulating creativity Developing future-oriented risk-taking projects Purchasing inexpensively*	Enhancing trust and cooperation Coordinating delivery and quality control'* Keeping corporate secrets *Implementing contracts flexibly* *Maintaining strict standards and tough relationship'*
Low Performance (Third and Fourth in Ranking)	Acquiring technological information Obtaining business, owing to long-term mutual experience Enhancing trust and cooperation Coordinating delivery and quality control Receiving stable supplies Receiving services to satisfy needs Keeping corporate secrets Stimulating creativity Developing future-oriented risk-taking projects *Implementing strategic pricing* *Purchasing inexpensively* *Implementing contracts flexibly* *Maintaining strict standards and tough relationship*	Coordinating delivery and quality control	Keeping corporate secrets *Altering procurement sources flexibly* *Maintaining strict standards and tough relationship*	Acquiring technological information Receiving stable supplies Receiving services to satisfy needs Stimulating creativity Developing future-oriented risk-taking projects *Creating competitive environment by introducing another company* *Purchasing inexpensively* *Altering procurement sources flexibly*

Note: This table is based on mean values in Tables 4.1 and 4.2.
Regular letters indicate flexible-synergy effects; *italic letters* indicate market-like effects.
* The top position is shared between horizontal and vertical interaction.
'* The top position is shared between vertical and internalized interaction.
' The top position is shared between horizontal and internalized interaction.

ing indicator of flexible-synergy effects. This evidence seems to clearly suggest that in the procurement of parts and materials, vertically- and horizontally-related companies mix cooperation and competition and create dynamic interaction. This is why in Table 3.7 these two types of interaction are highlighted with light shading as good examples of CCC interaction. This also seems to explain why 81.0% of procurement in parts and materials was transacted through long-term relations.

Ways to mix these types of effects, however, differ by partner. By comparing indicators in the category of top performance, it again becomes very clear that vertically-related suppliers are better at generating flexible-synergy effects, while horizontally-related suppliers were more effective at generating market-like effects.

Since vertically-related suppliers tended to be inferior to horizontally-related suppliers and intrafirm supply units in many aspects, especially in technological capability, they have had to outperform these technologically capable competitors. Hence, they have attempted to generate comparative advantages in daily operations rather than technological capability. Many of the flexible-synergy effects are targeted at reducing the transaction costs of operations by exchanging know-how with each other and satisfying mutual needs and expectations. Vertically-related suppliers provided parts and materials with most inexpensive prices, and even responded well to the extra demands of semiconductor companies. They even performed the best at stimulating creativity and taking future-oriented risk-taking activities, since these activities were the life line of their continued relations. No other partners could provide as many high level effects, although these companies were not very effective at keeping corporate secrets and meeting strict standards and tough relationships. Altering procurement sources flexibly was not a workable competition-generating measure for vertically-related customers, due to the stronger sense of cooperation they required. Basically, they tried to maintain their long-term relations by strongly appealing to the continuity of human relations and to a sense of cooperation and trust. Hence, they developed the human-relations oriented logic of continuity with semiconductor manufacturers.

Horizontally-related suppliers were also good, but different, partners for cooperation-oriented CCC interaction. Semiconductor manufacturers implemented especially stringent competition-generating measures. And these suppliers responded very effectively to strict standards and tough relations, strategic pricing, and even measures that resulted in low prices of parts and materials. To them, responding to cooperation-promoting measures was at an intermediate level. The primary importance of their behavior lay in performing the best, responding to such stringent competition-generating measures. They had a slightly lower need to emphasize cooperation than did vertically-related suppliers, and they tried to base their long-term relations more on the performance-oriented logic of continuity.

One interesting finding is that both vertically- and horizontally-related suppliers responded effectively to cutting production costs and giving lower prices than suppliers in the spot market or intrafirm supply units. Vertically-related suppliers that faced fewer competition-generating measures even performed as well as horizontally-related suppliers in terms of reducing prices. This was due to their strong sense of cooperation based on the human-relations-oriented logic of continuity, while horizontally-related suppliers responded better to competition-generating measures, basing their judgment on the performance-oriented logic of continuity. The two different approaches within CCC governance achieved a similar result.

The fact that vertically- and horizontally-related suppliers could give the lowest prices suggests that market governance may not have been the most effective institutional arrangement for generating the lowest price and creating allocative efficiency. A combination of measures involving both allocative and non-allocative efficiency may have become more efficient for allocating resources. For partners who emphasized performance more than cooperation, after developing some basic long-term cooperative relationship, the stringent application of competition-generating measures worked effectively to reduce prices. In contrast, for highly cooperation-oriented companies with fewer resources, a more stringent implementation of cooperation-promoting measures with a limited degree of competition-generating measures worked very effectively to reduce prices.

With regard to the latter case, some may argue that power asymmetry allowed the larger companies to exploit smaller ones. However, there was hardly any difference between the scores of vertically- and horizontally-related suppliers in the indicators of Purchasing Inexpensively. No evidence of exploitation over the price exists. It is true that small- and medium-sized companies often faced tough conditions. In the 1950s and 1960s, relationships between large- and small-sized companies in Japan were highly exploitative. But semiconductor manufacturers incurred enormous costs to build the capabilities of these companies. This help enabled vertically-related suppliers to equip themselves with internationally competitive technological and managerial capabilities. In some cases, semiconductor manufacturers grew dependent on their expertise. And even now, they continue to receive enormous benefits from semiconductor manufacturers, and mutual commitment to long-term relations ensures their survival. As the lists of flexible-synergy and market-like effects show, semiconductor manufacturers, in return, have also enjoyed enormous benefits. Despite the power asymmetry, these findings suggest that their shared goals and interests have been similar, and gains have been mutual. An alternative explanation could be that, given differences in resource endowments, both partners engaged in cooperative bargaining under power asymmetry and pursued the maximization of mutual gains.

In contrast with vertically- and horizontally-related suppliers, intrafirm supply units and suppliers in the spot market could not provide such well-balanced benefits. Surprisingly, intrafirm relations were not as effective as CCC interaction in generating both flexible-synergy and market-like effects. Semiconductor divisions and intrafirm supply units kept corporate secrets better, nurtured a stronger sense of trust and cooperation, and coordinated delivery and quality control better. But as Table 4.3 shows, performance within these relationship was more often identified as low rather than as high or intermediate. The most devastating problems were their inability to alter procurement sources flexibly and their difficulty in creating a competitive environment by introducing other suppliers. Consequently, prices became higher than those of vertically- and horizontally-related suppliers. The lack of flexibility found in relationships with intrafirm suppliers is generally perceived as a characteristic of hierarchical governance. The findings clearly suggest that semiconductor divisions and intrafirm supply units are actors harmonized under hierarchical governance.

The interaction between semiconductor companies and suppliers in the spot market was generally the weakest in terms of generating both flexible-synergy and market-like effects. However, these relationships could easily enable the introduction of another supplier, the creation of a competitive environment, and even the flexible alteration of procurement sources. These characteristics resemble the generally-accepted qualities of market governance. The description above fits well with the predicted characteristics of governance indicated in Table 3.7.

Hence, CCC interaction in the procurement of parts and materials resulted from a well-balanced combination of flexible-synergy and market-like effects. Although differences can be found between vertically- and horizontally-related suppliers in terms of the orientation of the logic of continuity and the combination of the two types of effects, this delicate mixture of cooperation and competition is the basic characteristic of competitive-cum-cooperative governance.

Does the pattern of CCC interaction found in parts and materials hold for manufacturing equipment? Does CCC interaction generate flexible synergy benefits for these partners in the same way? Does it provide the same benefits with regard to market-like effects? The next section answers these questions.

B.   Procurement of Manufacturing Equipment

In the previous chapter, I found that oligopolistic and technologically-advanced horizontally-related equipment suppliers tended to have power-symmetric relations with semiconductor manufacturers. They shared technology, developed firm-specific equipment cooperatively, and dominated the equipment market (48.8%). Why were they so strong? Was it solely because of their technology?

Vertically-related suppliers also engaged in developing equipment in less sophisticated technology areas, but they faced suppressive control exercised by semiconductor manufacturers (Table 3.7). In the previous chapter I raised the question whether vertically-related suppliers could really develop cooperation-oriented CCC interaction under suppressive conditions. But they did supply 25.5 % of all manufacturing equipment. Why were they in demand?

Intrafirm procurement had been vital for developing forefront technology, while spot market purchases were quite insignificant. Why were these types of procurement minor in manufacturing equipment? I shall again analyze the importance of these types of procurement by examining their flexible-synergy and market-like effects.

1.   Flexible-synergy Effects

a.   *Benefits from Long-term Mutual Accumulation*

The benefits gained by semiconductor companies from long-term mutual accumulation attained their highest score at 3.20, a medium+ level, as shown in Table 4.4; this suggests that the procurement of manufacturing equipment involved fewer benefits from cumulative interaction than was gained in the procurement of parts and materials. This phenomenon seems to be simply because manufacturing equipment was not purchased frequently, and did not require continuous interaction in negotiations and improvements. Despite these facts, long-term relations and a sense of trust and cooperation have been considered important. Horizontally- and vertically-related suppliers recognized the contributions of long-term relations to obtaining business (at a medium+ level). Intrafirm manufacturing equipment units also showed a slightly lower score than vertically- and horizontally-related suppliers (at a medium+ level). And both intrafirm manufacturing units and vertically-related suppliers recognized that long-term relations could enhance a sense of cooperation and trust (at a medium+ level), while horizontally-related suppliers showed a slightly lower tendency to develop these qualities (at a medium level). This was because all types of partners besides companies in the spot market shared the goals of mutual prosperity, technology, and speedy and special development of equipment with semiconductor manufacturers (see Common Goals and Interests in Table 4.4).

Both vertically-and horizontally-related suppliers contribute additional benefits to semiconductor manufacturers' operations. Because of their oligopolistic positions in equipment markets, horizontally-related suppliers were a little more helpful in providing both technological and market information to semiconductor manufacturers (at a medium+ level) than any other type of partner. Vertically-related suppliers were helpful in providing technological information (at a medium+ level), but less so in providing market information (at a medium level).

The crucial contribution of horizontally- and vertically-related suppliers was the joint development of firm-specific equipment (see Common Goals and Interests in Table 4.4). Once the practices of joint development were well established, the accumulated past technology started to restrict or guide the development of future equipment. Changing equipment manufacturer would have meant the loss of accumulated technology and information, as well as increased uncertainty regarding technological quality and the yielding rate of semiconductors. Besides, to change equipment or even a part, semiconductor companies would have needed to obtain permission from customers, since changes could have influenced the performance of semiconductors.[11] Given the firm-specific nature of technology, cooperative relations between semiconductor and equipment manufacturers became essential. In other words, technology was the most important reason for maintaining a long-term relationship.

Even for horizontally-related suppliers in the oligopolistic market, maintaining technological leadership required developing a good network to obtain information on technology and product markets and find opportunities for joint developments of manufacturing equipment. Long-term cooperation was essential to remain strong in this field. Some equipment manufacturers, however, were so monopolistic that they were the only company dealing with the most of the big ten semiconductor manufacturers. In these cases,

---

[11] Interview with Company V manager.

Table 4.4 Flexible-synergy Effects by Type of Interaction in Procurement of
Manufacturing Equipment

BENEFITS	TYPE OF INTERACTION			
	SPOT INTERACTION (N = 5)	HORIZONTAL INTERACTION (N = 5)	VERTICAL INTERACTION (N = 5)	INTERNALIZED INTERACTION (N = 5)
A. From Long-term Mutual Accumulation				
Acquiring Technological Information*	2.80	3.20	3.00	2.75
Acquiring Product-market Information*	2.80	3.20	2.80	2.50
Obtaining Business, Owing to Long-term Mutual Experience*	2.00	3.20	3.20	3.00
Enhancing Trust and Cooperation*	2.20	2.80	3.00	3.00
B. From Satisfying Mutual Expectations				
Coordinating Delivery and Quality Control*	2.40	3.00	3.60	3.00
Coordinating the Development of New Equipment*	2.00	3.40	3.20	3.25
Coordinating the Maintenance of Equipment*	2.20	3.20	3.20	3.00
Receiving Stable Supplies*	2.20	3.00	3.40	3.00
Receiving Services to Satisfy Needs*	2.60	3.20	3.40	2.75
Keeping Corporate Secrets*	2.00	2.60	3.20	4.00
C. From Future-oriented Risk-taking Activities				
Stimulating Creativity*	2.40	3.20	2.60	3.25
Developing Future-oriented Risk-taking Projects*	2.00	2.80	3.00	3.25
D. Common Goals and Interests				
	- Mutual profit	- Mutual profit - Mutual prosperity - Technology - Speedy and special development of equipment	- Mutual profit - Mutual prosperity - Technology - Speedy and special development of equipment	- Mutual profit - Mutual prosperity - Technology - Speedy and special development of equipment
	- Quick purchase	- Information	- Continuous dynamic interaction	- Development of frontier technology

Note: *1.0 means very weak and 5.0 means very strong. Scores are categorized as follows: very weak (mean value = 1.0 -
1.4); weak (1.5 - 1.9); weak+ (2.0 - 2.4); medium (2.5 - 2.9); medium+ (3.0 - 3.4); strong (3.5 - 3.9); strong+ (4.0 –
4.4); and very strong (4.5 - 5.0).

it became very difficult for suppliers to talk about interdependence, trust and cooperation with a particular semiconductor manufacturer. Unlike vertically-related suppliers who tend to develop cooperation in extended areas, their cooperation remains in their specific areas of technological expertise. Hence, semiconductor manufacturers' interaction with horizontally-related suppliers became slightly more restricted and performance-oriented.

Less than 20% of vertically-related equipment manufacturers were partly owned by semiconductor companies. However, most did show very strong loyalty to semiconductor manufacturers and they enjoyed diverse benefits from cooperation.[12] This was the only type of partner with which semiconductor companies would like to achieve dynamic mutual development (see Common Goals and Interests in Table 4.4).

For example, in an early period of the semiconductor industry, Company E, a very small company with only about fifteen workers, agreed to venture into the manufacturing of tester equipment when Company Q could not find a risk-taking company to do it. Company Q provided financing, machines, engineers, and training, until the former became financially and technologically competent enough to develop any new product. At the time of the survey, it allowed Company E to receive its detailed technological and market information, use its training facilities, enter freely into its factories, and even obtain help in its production planning. One example demonstrates the intensity of their interaction. When they started a new project, four engineers of Company Q discussed issues with engineers of Company E three times a week for over a month. Company Q sent one engineer to its partner for one week to provide detailed engineering guidance, and it supplied necessary parts for producing a testing system, sometimes for free. Since Company E was a small company then, it lacked the ability to import some special integrated circuits quickly. Company Q used its foreign affiliates to purchase these parts for Company E. Once the development was successful, Company E was guaranteed a sufficient number of orders for a certain period of time. Success in past projects then led to more development projects, securing the continuous operation of the company and expanding operations significantly.

Company E began producing much more cost-effectively and responding more flexibility to company Q's demands than another company that Company Q partly owns, Company G. This disparity was due to the fact that Company G had a similar salary scheme, work system, and calendar holi-

---

[12] Interview with Company Q manager.

days and vacations as Company Q. They did not work when Company Q was in holiday and their overtime pay was much higher than that of Company E. In contrast, Company E responded to Company Q's demands flexibly, by doing things like postponing a vacation period. In addition, being a small- or medium-sized company, workers' wages were much lower than the salary scheme of Japan's leading semiconductor manufacturer. The accumulation of past experiences and practices that resulted from their joint development project enhanced their sense of trust and cooperation. Hence, Company Q ended up relying more on Company E, and these partners continued to engage in new projects.

The benefits and business opportunities that Company E enjoyed were enormous, though the organization became quite vulnerable to the demands of Company Q. Although the president of the smaller company at an early period earned a much lower annual income than the interviewee, who was a former top-level manager of Company Q, he later became the highest tax payer in his region. His risk-taking behavior successfully turned this rare opportunity into a successful long-term relationship. Despite asymmetric power relations, interaction was mutual and dynamic with a very strong sense of human bonding. This evidence clearly suggests that semiconductor companies and vertically-related suppliers mutually shared goals and interests and based their judgments on cooperative bargaining, which is a necessary condition for CCC interaction.

b. *Benefits from Satisfying Mutual Expectations*

Six indicators were related to the benefits semiconductor companies gained from satisfying mutual expectations with each type of partner: three dealt with the smoothness of coordination; and the other three dealt with equipment manufacturers' compliance with semiconductor companies' demands.

Since the sales of manufacturing equipment were based on developing firm-specific technology, smooth coordination for developing new equipment was essential to all types of partners (at a medium+ level in Table 4.4) except suppliers in the spot market, who sold even firm-specifically developed equipment as standardized products. But differences between types of partners became apparent, after they developed new equipment.

The maintenance of equipment and diverse related services were an essential part of after-sales activities. Both vertically- and horizontally-related suppliers performed a little better (at a medium+ level) than intrafirm equipment manufacturing units and much better than suppliers in the spot market. And both vertically- and horizontally-related suppliers performed better (at a medium+ level) than any other type of partner in providing diverse services to satisfy semiconductor companies' needs. The President of Shinkawa, the world's number-two automatic bonding machine producer in 1989, said that his company tried not to develop its product market beyond its service capability. Once it sold a machine to a customer, it did its best not to lose the trust of customers by providing high quality services and maintenance (Toyo Keizai, 1982:53). This medium-sized vertically-related company worked to establish trust by providing diverse services and the best maintenance. This practice suggests why vertically-related suppliers performed a little better than other types of partners.

Vertically-related suppliers' strength does not end only in maintenance-related activities. They were also ranked as the best performers in terms of coordinating delivery and quality control (at a strong level) and mutually working to keep a stable supply of equipment (at a medium level). For technologically weak vertically-related suppliers, organizational coordination over delivery, quality control, and providing a stable supply of equipment became important comparative advantages. As previously mentioned in the case of Company E, meeting a tough delivery schedule was one strong advantage that many other types of partners did not provide.

It seems that power asymmetry generated conditions under which a big semiconductor manufacturer could pressure less powerful equipment manufacturers to comply with their urgent needs. Vertically-related suppliers faced especially difficult conditions, and they had to take risks in order to hang on to semiconductor manufacturers. But their relationships were not necessarily exploitative. For an industry in which the product market fluctuates severely, a stabilizing mechanism becomes necessary. When demand for semiconductors was rapidly increasing, it became necessary for semiconductor manufacturers to ensure the supply of equipment. Vertically-related suppliers responded better to their needs in such conditions. But in slack times, semiconductor manufacturers kept ordering extra equipment, so that vertically-related suppliers could survive and did not lose accumulated technology by being forced to lay off workers. Such reciprocal behavior was an important incentive for both to maintain a long-term relationship (Nikkei Business, 1984a).

Again, suppliers in the spot market were not good at satisfying the needs of customers and keeping a stable supply. But when standardized equipment was offered very inexpensively, or a highly-sophisticated machine was needed as a one-time purchase, semiconductor companies purchased equipment from

---

[13] Interview with Company X manager.

the spot market.[13] These purchases were the reasons why the percentage of manufacturing equipment procurement made in the spot market was a bit higher than the percentage of parts and materials.

Keeping corporate secrets is important, especially when the equipment produced is highly firm-specific. Intrafirm equipment manufacturing units kept very tight control over corporate secrets (at a strong+ level). They received the highest score involved in the procurement of manufacturing equipment. One strange finding is that horizontally-related suppliers showed a very low level (a medium level) in comparison with the procurement of parts and materials. This evidence contradicts my earlier statement that corporate secrets are well secured in oligopolistic companies that deal with multiple semiconductor manufacturers in a highly technologically sophisticated area. A former top manager of Company Q explained that less important technologies usually involved verbal agreements in place of written secrecy contracts. Due to the oligopolistic conditions of some equipment markets, one company actually catered to the needs of many companies. A less-sensitive technology developed for one company could be easily made available to other customers with some alteration, raising the capability of competitors. Such practices generated technological synergy, while technological dissemination stimulated market competition. This may explain the lower score received by horizontally-related suppliers.

c.   Benefits from Future-oriented Risk-taking Activities

Future-oriented risk-taking activities were indispensable for developing grounds for continued long-term relations. Due to their technological capability, horizontally-related suppliers and intrafirm equipment manufacturing units stimulated creativity equally well (at a medium+ level in Table 4.4), while vertically-related suppliers and those in the spot market did not stimulate creativity so well (at a medium level). In contrast, intrafirm equipment manufacturing units developed future-oriented risk-taking projects more than any other partner (at a medium+ level). Vertically-related suppliers (at a medium+ level) and horizontally-related suppliers (at a medium level) then follow. It was surprising to find that highly-stimulating horizontally-related suppliers were not participating well in future-oriented risk-taking projects, while less-stimulating vertically-related ones were more willing to develop projects.

The best type of partner for turning new stimulating ideas into future-oriented risk-taking activities were intrafirm equipment manufacturing units, because one of their shared goals and interests was developing frontier technology (see Table 4.4). These units were the natural choice for semiconductor divisions for several reasons: (1) Intrafirm equipment manufacturing units and research laboratories were equipped with a highly sophisticated level of technology; (2) it was easier to keep corporate secrets inside a company; (3) developing equipment inside a company was inexpensive; and (4) costs could be written off as research costs. In developing a high-risk machine, semi-conductor manufacturers, for example, may have ended up making five different machines, out of which only one may have been used. To develop five machines outside of a company costs more than doing so inside the company. But a more important reason was that purchasing five machines from outside the company would make a part of the taxable assets of a company, while internal development reduced corporate taxes because it could be written off as a research cost.[14]

Oligopolistic horizontally-related suppliers, despite their intensive R&D and many good stimulating ideas for future equipment, were not necessarily willing to cooperate with semiconductor manufacturers for future technological development. They were willing to develop manufacturing equipment jointly with semiconductor manufacturers, but not to engage in advanced research activities. They usually made independent decisions and carried their own risks, unless it became necessary to share costs and technology.

In contrast, vertically-related suppliers were relatively weak in developing stimulating technological ideas, but they were willing to engage in future-oriented risk-taking projects. Sometimes, semiconductor manufacturers shared costs and risks for developing new equipment with vertically related suppliers, which encouraged the latter to take more risks. Practices like these were not only done to promote the innovation of new equipment, but also they kept vertically-related suppliers technologically capable. Their failure to comply with the demand clearly means the violation of satisfying mutual expectations (Aida, 1992:365), and may have threatened their future long-term relations. In other words, for vertically-related suppliers, developing a path for future business transactions in cooperation with semiconductor companies was the only way for them to survive, while the superior technology of horizontally-related suppliers attracted customers on its own. Although suppliers in the spot market may have had some interesting ideas for future technology, they were the least reliable partners for cooperation (at a level of weak+).

Thus, I found that long-term mutual accumulation in CCC interaction

---

[14] Interview with Company Q manager.

benefits semiconductor manufacturers by helping them to acquire technological information and generate business opportunities, while the emphasis
on satisfying mutual expectations in CCC interaction helped them to better
coordinate the maintenance of equipment and the provision of services for
satisfying their needs. CCC interaction worked well to help suppliers satisfy
the extra needs of semiconductor companies, but future-oriented risk-taking
activities were achieved in an intermediate level. In terms of the number of
indicators, CCC interaction showed its strength in four items, internalized
interaction in two items, and spot interaction in none. CCC interaction was
still important, but it was not as conspicuous as it was in the area of parts and
materials acquisition. This was because the procurement of manufacturing
equipment is less frequent than that of parts and materials.

Of the different types of partner, vertically-related suppliers were the best
sole performers in three out of twelve indicators, horizontally-related suppliers in two, both of these types of suppliers tied in two categories, and intrafirm
equipment manufacturing units were the best in one. CCC interaction topped
seven indicators out of twelve, suggesting that it was still a dynamic source
of flexible-synergy effects even in a less interactive context.

2.   Market-like Effects

a.   Benefits from Market-linked Adjustments

Vertically-related suppliers responded to strategic pricing and introducing
another supplier better than other types of partners (at a medium+ level in
Table 4.5), while horizontally-related suppliers and intrafirm equipment
manufacturing units responded a little less effectively (at a medium level).
Such measures were barely effective with suppliers in the spot market (at a
weak level).

Since manufacturing equipment was not procured constantly, implementing a strategic pricing was not easy. Besides, the sophistication and firm-
specific nature of technology made it more difficult to introduce another
supplier to generate competition and create a market-like environment. But
given the less sophisticated, yet firm-specific, nature of technology that vertically-related suppliers handled, implementing strategic pricing and introducing another supplier may not have been that difficult for semiconductor
companies.

Introducing another supplier can generate diverse impacts on the operations of vertically-related suppliers. For example, Company L was a subsidiary of Company Q with a strong technological background. In the past, it
catered only to company Q's needs. When Company Q became dissatisfied

Table 4.5 Market-like Effects by Type of Interaction in Procurement of Manufacturing Equipment

	TYPE OF INTERACTION			
BENEFITS	SPOT INTERACTION (N = 5)	HORIZONTAL INTERACTION (N = 5)	VERTICAL INTERACTION (N = 5)	INTERNALIZED INTERACTION (N = 5)
A. From Market-linked Adjustments				
Implementing Strategic Pricing* 1.80		2.60	3.00	2.75
Creating Competitive Environ- 2.00 ment by Introducing Another Company*		2.80	3.20	2.75
Purchasing Inexpensively* 1.80		2.40	2.60	2.50
Altering Procurement Sources 3.20 Flexibly*		3.20	2.80	2.50
B. From Resource Interdependence Adjustments				
Implementing Contracts Flexibly* 1.80		2.80	3.00	3.25
C. From Reevaluation Adjustments				
Maintaining Strict Standards 2.40 and Tough Relationship*		3.00	3.00	2.67

Note: * 1.0 means very weak, and 5.0 means very strong. Scores are categorized as follows: very weak (mean value = 1.0 - 1.4); weak (1.5 - 1.9); weak+ (2.0 - 2.4); medium (2.5 - 2.9); medium+ (3.0 - 3.4); strong (3.5 - 3.9); strong+ (4.0 - 4.4); and very strong (4.5 - 5.0).

with the quality of products, it reduced its purchases from the company and started purchasing from another supplier. The reduction of demand forced Company L to sell their products to other semiconductor manufacturers. Expanding its customers was not problematic, since their products involved less corporate secrecy. Introducing another supplier not only created competition between Company L and the new supplier, but also exposed Company L to many claims and pressures by other semiconductor manufacturers. The company came to recognize its need to introduce more advanced technology, so it made a cross-licensing agreement with a U.S. company. Its newly developed technological superiority helped the company gain its sales back from Company Q, and Company L again came to enjoy an oligopolistic market position.[15]

Technologically-advanced horizontally-related suppliers, however, could weaken the impact of semiconductor manufacturers and make the latter dependent on them. But in some cases, the dependence became so intolerable that the suppliers broke off the relationship. For example, Company R decided to break an oligopolistic condition generated by Company J and Company K. On orders placed by Company R two years in advance of purchase,

---

[15] Interview with Company Q manager.

both Company J and Company K failed to reach the desired targets for quality, delivery, services, and in particular, price discounts. Hence, Company R decided to nurture a small company by giving technology and financial aid, and succeeded in developing a highly quality-conscious well-reputed company. Losing some business, Company J responded by offering a lower price and regained some of its orders back. The strategy played by Company R worked as planned.[16] This incident shows again how an additional supplier could be introduced to create a competitive environment.

Unlike parts and materials suppliers, equipment manufacturing companies were not very responsive to pressures to reduce a price for a jointly developed equipment. All partners except suppliers in the spot market tended to give favorable prices to semiconductor manufacturers, though at quite low levels (at weak+ to medium levels). I found here that the spot market was not generating lower prices.

The spot market was attractive because semiconductor manufacturers could change equipment suppliers flexibly. One surprising finding is that it was equally easy to change suppliers when dealing with horizontally-related companies. Given the powerful oligopolistic positions of horizontally-related suppliers and slightly weaker sense of cooperation, as long as an alternative source of procurement was available, altering suppliers remained as a potent measure for enhancing the bargaining position of semiconductor companies. In contrast, because of a strong sense of cooperation and interdependence, semiconductor companies found it a bit more difficult to alter procurement sources flexibly with vertically-related suppliers and intrafirm equipment manufacturing units (at a medium level).

b.   Benefits from Resource Interdependen Adjustments and Reevaluation Adjustments

Machinery delivery dates were not as flexible as parts and materials. Instead, delivery dates indicated when a new machine is to be delivered, whether the completion of a new machine had to be speeded up or could be delayed. This type of negotiation was rather easy to do with intrafirm equipment manufacturing units and vertically-related suppliers having very close cooperative relations (at a medium+ level); it was more difficult with horizontally-related suppliers (at a medium level). Suppliers in the market were the least responsive to this type of demand.

---

[16] Interview with Company R manager.

To maintain strict standards and a tough relationship was one of the most important competition-generating measures for keeping pressure on interacting partners. Strict standards were especially important for procuring precision equipment. Again vertically- and horizontally-related suppliers responded better than any other type of partner (at a medium+ level). Horizontally- and vertically-related suppliers were preferred by semiconductor manufacturers, especially because of their capability to respond well to strict standards and tough negotiations. To implement such practices vis-á-vis intrafirm equipment manufacturing units or suppliers in the spot market was rather difficult. Several managers of semiconductor manufacturers[17] expressed difficulty in dealing with people in other divisions of the same company, especially when the development of forefront technology was involved. This may have been simply because a semiconductor division did not have sufficient expertise in developing frontier technology, and had to be dependent on intrafirm equipment manufacturing units.

Thus, in the procurement of manufacturing equipment, both vertically- and horizontally-related suppliers were effective in meeting strict standards and facing tough relationships. But due to difference in their technological capability, they behaved differently. Vertically-related suppliers were top sole performers in two measures of market-linked adjustments, while oligopolistic horizontally-related suppliers received top ratings in none, suggesting their stronger bargaining and less cooperative positions. Intrafirm equipment supply units performed the best in only flexible contracting. CCC interaction may have still been important, despite the generally less interactive environment. An analysis in the summary section provides a clearer picture.

3.   Summary: Cooperation-oriented CCC Interaction and Specialization

Since the purchases of semiconductor equipment tended to be infrequent, interaction between semiconductor companies and manufacturing equipment suppliers was less frequent than it was between the partners for procurement of parts and materials. But this statistics does not mean that long-term relations did not exist. As a matter of fact, interfirm relations in this functional area were more long-term oriented than in any other functional area, because many portions of manufacturing equipment were firm-specific, and because

---

[17] Interviews with managers of Companies Q, R and V.

they were developed jointly between semiconductor manufacturers and equipment manufacturers. Once the practice of joint development was well established, the accumulated past technology bound the development of future equipment. And changing equipment or even a part required the permission of major customers and sometimes involved unpredictable difficulty. Hence, technology became the most important reason for maintaining long-term relations.

Table 4.6 shows the performances of each type of interaction. By looking at the number of indicators in the category of low performance, the table clearly shows that partners in spot interaction did not contribute to flexible-synergy effects. Eleven out of twelve indicators related to flexible-synergy effects are in the category. Internalized interaction had five indicators out of twelve in the same category. It was relatively weak in terms of providing extra services and technological and market information, generating continuous business, and coordinating the maintenance of equipment.

In contrast, horizontally-related companies had only three indicators in the category of low performance, showing their weakness in keeping corporate secrets, developing future-oriented risk-taking activities, and enhancing trust and cooperation. Vertically-related companies had only one indicator, suggesting their weakness in stimulating creativity. These findings suggest that CCC interaction was still at the core of generating flexible-synergy effects, even in this less interactive environment.

A comparison of indicators in the category of top performance suggests that there has been a clear division of labor among all types of partners other than suppliers in the spot market. Vertically-related companies specialized in less technologically sophisticated machinery, while horizontally-related suppliers sold highly technologically sophisticated machines. Intrafirm equipment manufacturing units specialized in developing forefront technology.

Vertically-related suppliers, being weaker in technological capability, performed the best by generating flexible-synergy effects in areas that were closely related to daily operations. They made use of cumulative experience and satisfied the extra needs of semiconductor companies. They also tried to maintain long-term relations by engaging in future-oriented risk-taking activities, though at an intermediate level. Their dynamic performance was strengthened by the implementation of less threatening competition-generating measures such as introducing other suppliers, tough standards, and strategic pricing.

Again, the technological vulnerability of vertically-related suppliers was a reason for responding well to both flexible-synergy and market-like effects, especially as these existed under the strong control of semiconductor

Table 4.6 Performance by Type of Interaction in Procurement of Manufacturing Equipment

PERFORMANCE	TYPE OF INTERACTION			
	SPOT INTERACTION	HORIZONTAL INTERACTION	VERTICAL INTERACTION	INTERNALIZED INTERACTION
Top Performance (Top in Ranking)	*Altering procurement sources flexibly*[‡]	Acquiring technological information Acquiring product-market information Coordinating the maintenance of equipment Obtaining business, owing to long-term mutual experience* Stimulating creativity[†] *Altering procurement sources flexibly*[‡] Maintaining strict standards and tough relationship*	Enhancing trust and cooperation[%] Obtaining business, owing to long-term mutual experience* Coordinating delivery and quality control Coordinating the maintenance of equipment Receiving stable supplies Receiving services to satisfy needs Creating competitive environment by introducing another company Maintaining strict standards and tough relationship* Implementing strategic pricing	Enhancing trust and cooperation[%] Keeping corporate secrets Stimulating creativity[†] Developing future-oriented risk-taking projects *Implementing contracts flexibly*
Low Performance (Third and Fourth in Ranking)	Acquiring technological information Obtaining business, owing to long-term mutual experience Enhancing trust and cooperation Coordinating delivery and quality control Coordinating the development of new equipment Coordinating the maintenance of equipment Receiving stable supplies Receiving services to satisfy needs Keeping corporate secrets Stimulating creativity Developing future-oriented risk-taking projects Implementing strategic pricing Creating competitive environment by introducing another company Purchasing inexpensively Implementing contracts flexibly *Maintaining strict standards and tough relationship*	Enhancing trust and cooperation Keeping corporate secrets Developing future-oriented risk-taking projects *Implementing contracts flexibly*	Stimulating creativity *Altering procurement sources flexibly*	Acquiring technological information Acquiring product-market information Obtaining business, owing to long-term mutual experience Coordinating the maintenance of equipment Receiving services to satisfy needs *Altering procurement sources flexibly* *Maintaining strict standards and tough relationship*

Note: This table is based on mean values in Tables 4.4 and 4.5. Regular letters indicate flexible-synergy effects; *italic letters* indicate market-like effects.
* The top position is shared between horizontal and vertical interaction; [%] The top position is shared between vertical and internalized interaction; [†] The top position is shared between horizontal and internalized interaction; and [‡] The top position is shared between horizontal and spot interaction.

companies (see Table 3.7). Despite asymmetric power relations, interaction was mutual and dynamic with a very strong sense of human bonding. Long-term relations enhanced trust and cooperation, and vertically-related companies strongly followed the human-relations-oriented logic of continuity. They mutually shared goals and interests and based their judgments on cooperative bargaining. They clearly set a pattern of cooperation-oriented CCC interaction.

In contrast, horizontally-related suppliers offered semiconductor manufacturers attractive benefits in the areas of technology and market information and stimulation of ideas. They also performed well in terms of coordinating the maintenance of equipment, and their technological linkages generate more business. They were able to provide these benefits because they jointly developed highly sophisticated manufacturing equipment. Despite doing so, horizontally-related suppliers were a little less willing to take future-oriented risk. They tried to maintain their technological superiority by strongly relying on their internal capabilities, and limited the areas of cooperation to the joint development of highly sophisticated firm-specific equipment and information exchange. As shown in Table 3.7, horizontally-related suppliers in manufacturing equipment were the least cooperative type of partner for semiconductor companies, besides suppliers in the spot market. Evidence for this conclusion is also shown in Table 4.6, for one low performance indicator was enhancing trust and cooperation. Hence, despite their capability, they generated flexible-synergy and market-like effects only in a limited area, making them a weak partner under cooperation-oriented CCC governance.

Given their commitment to long-term relations based on technological reasons, they clearly based their behavior on the performance-oriented logic of continuity. But their lack of commitment to relations other than technological linkages made them different from other types of partners. Technology did bind partnerships, therefore as long as this linkage existed, commitment to long-term relations existed. Such relationships were sustained because of oligopolistic market conditions. When their positions became technologically weaker, or became too strong in a way that prevented the mutual sharing of goals and interests, as in the case of Company R vis-á-vis Company J and Company K, drastic changes were likely to happen. Such environments allowed both partners to accept the severest measure of flexibly altering procurement sources, which could not be easily applied to vertically-related suppliers or intrafirm equipment manufacturing units. Another reason for this competition-generating measure was to strengthen the bargaining position of semiconductor companies to powerful horizontally-related suppliers. Hence, the characteristics of horizontal interaction suggest that the logic of continuity was weak. The interaction may possibly shift from cooperation-oriented CCC interaction to competition-oriented CCC, once the market becomes less oligopolistic. But it will never become spot interaction, since cooperation is always necessary for developing firm-specific technology. This is why horizontal interaction in manufacturing equipment has been closer to competition-oriented CCC governance than to a cooperation-oriented one as shown in Table 3.7. I have found evidence of potentially unstable cooperation-oriented CCC interaction.

Then who engaged in developing forefront manufacturing equipment in cooperation with semiconductor divisions? This was the role of intrafirm manufacturing equipment units. Their top performance was in keeping tight technology secrecy, stimulating innovative ideas, and engaging in future-oriented risk-taking activities. They had fewer advantages in creating both flexible-synergy and market-like effects. Hence, it was difficult for them to compete against horizontally-related companies in supplying highly advanced manufacturing equipment and against vertically-related suppliers in supplying less advanced manufacturing equipment. Their strength lay in high technological capability, strong cooperation and coordination, and corporate secrecy maintenance. Hierarchical governance, well supported by favorable tax conditions and cost factors, provided the best environment for forefront R&D activities.

Suppliers in the spot market could be the best type of partner for semiconductor companies when they needed to alter partners frequently. Clearly, they showed characteristics of actors in market governance. They were also a good source of market information, but their inability to generate flexible-synergy and market-like effects made them critically disadvantaged in competitions against other partners in regular transactions.

Cooperation-oriented CCC interaction was found prevalent in the functional areas of both parts/material and manufacturing equipment procurement. But the environments in which interaction took place were quite different. Then, how was CCC interaction different in these two functional areas?

C.   Comparison of CCC Interaction in the Procurement of Parts/Materials and of Equipment

Semiconductor companies operated, regarding the procurement of parts/materials and equipment, quite differently. Parts/material purchases were a part of daily operations that required intensive cooperation and coordination over delivery and quality control. In contrast, equipment purchases did not occur frequently. Despite this difference, I found CCC interaction to have been important in both functional areas, though it was less so in the case of manufacturing equipment.

There were two major factors that greatly differentiated the content of CCC interaction. One was the difference in reasons for sustaining cooperation. Parts/material relations were for maintaining supply linkages, while equipment relations were for keeping technological linkages. The other influential factors were the characteristics of partners, which were associated with their positions in the market. The equipment market was oligopolistic, while parts/material markets were more, but not severely, competitive.

Reasons for sustaining cooperation were different in the two functional areas. Semiconductors are highly sensitive devices, reacting to any little change in factory environments, machinery, and the quality of parts and materials. Thus, coordination over the quality of parts and materials tended to be especially important. To cope with problems, parts/material suppliers and semiconductor manufacturers have established a computer system that enables them to trace every small lot of products and delivery. They interacted quite frequently, and developed strong cooperative relations. Long-term relations were a part of daily operations based on supply linkages.

In contrast, manufacturing equipment tended not to be purchased frequently. When semiconductor companies expanded a production line or built a new factory, they invested a large sum of money in equipment. But purchases were not continuous. This does not mean that long-term relations did not exist. Equipment was firm-specific, jointly developed by equipment and semiconductor companies, which limited future development and forced them to develop cooperative long-term relations. Long-term relations were based on technological linkages.

These differences influenced CCC interaction. Table 4.7, derived from Tables 4.3 and 4.6, compares CCC benefits by type of interaction in two different functional areas. It is quite interesting to find that both vertically- and horizontally-related equipment suppliers had a stronger sense of generating business based on long-term relations than did parts/material suppliers. Technological linkages cultivated a firmer sense of mutual accumulation than did supply linkages. Long-term mutual accumulation was more valued in the equipment area.

Since parts/material procurement was based more on daily operations, their advantages were derived from satisfying the extra needs of semiconductor companies by complying with demands for urgent deliveries and quick changes. In contrast, the advantages of equipment suppliers were not only in

Table 4.7 Comparison of CCC Benefits by Type of Interaction in Procurement of Parts and Materials and of Manufacturing Equipment

BENEFITS	PROCUREMENT OF PARTS AND MATERIALS			PROCUREMENT OF MANUFACTURING EQUIPMENT		
	FROM VERTICAL AND HORIZONTAL INTERACTION	FROM VERTICAL INTERACTION	FROM HORIZONTAL INTERACTION	FROM VERTICAL AND HORIZONTAL INTERACTION	FROM VERTICAL INTERACTION	FROM HORIZONTAL INTERACTION
A. Flexible-synergy Effects						
1. Benefits from Long-term Mutual Accumulation	Acquiring technological information	Obtaining business, owing to long-term mutual experience		Acquiring technological information Obtaining business, owing to long-term mutual experience	Enhancing trust and cooperation[b]	Acquiring product-market information
2. Benefits from Satisfying Mutual Expectations	Providing stable supplies Receiving services to satisfy needs	Coordinating delivery and quality control[b]		Coordinating the maintenance of equipment	Coordinating delivery and quality control Receiving stable supplies Receiving services to satisfy needs	
3. Benefits from Future-oriented Risk-taking Activities	Stimulating creativity Developing future-oriented risk-taking projects			None		Stimulating creativity[a]
B. Market-like Effects						
1. Benefits from Market-linked Adjustments	*Purchasing inexpensively[*]*		*Implementing strategic pricing*	*None*	*Implementing strategic pricing Creating competitive environment by introducing another company*	*Alternating procurement sources flexibly[s]*
2. Benefits from Resource Interdependence Adjustments	*None*			*None*		
3. Benefits from Reevaluation Adjustments			*Maintaining strict standards and tough relationship[a]*	*Maintaining strict standards and tough relationship[a]*		

Note: Regular letters indicate flexible-synergy effects; *italic letters* indicate market-like effects.
[*] The top position is shared between horizontal and vertical and internalized interaction; [b] The top position is shared between vertical and internalized interaction; [a] The top position is shared between horizontal and internalized interaction; and [s] The top position is shared between horizontal and spot interaction.

the technology itself, but were also related to good performance in terms of coordinating the maintenance of equipment. Parts/material partnerships were based more on daily interaction, while partnerships with equipment manufacturers were based on the maintenance of conditions. If conditions were good, then interaction was not necessary. Interaction became secondary.

This difference also influenced the effectiveness of particular types of competition-generating measures. In parts and materials, price adjustments were based primarily on periodic changes in the market, while equipment prices were often determined by a one-time cost of production and usually did not involve any change. Hence, parts and materials acquisition required more frequent interaction for adjustments than equipment procurement. For parts/material suppliers, competition-generating measures for market-linked adjustments were severe, but those measures functioned to cement future cooperative relations. Since reducing the price of jointly developed equipment was rather difficult, the focus of competition-generating measures was placed on maintaining strict standards and a tough relationship. This difficulty in reducing prices may have been one of crucial reasons why semiconductor divisions developed forefront manufacturing equipment by relying on internal equipment manufacturing units, where they could make trials and errors without involving exorbitant costs. Market-linked adjustments were found more effective when daily interaction and response to the market were possible, as in the case of parts and materials supply partnerships.

For parts/material suppliers, future survival was based on developing future-oriented risk-taking projects simply because they needed to develop future potential business in order to maintain long-term relations. In a sense, the logic of continuity was fulfilled by both vertically- and horizontally-related companies in the procurement of parts and materials. Hence, they were good examples of cooperation-oriented CCC interaction.

Horizontally-related equipment suppliers failed to fulfill one of the crucial criteria for the logic of continuity. They were less willing to engage in future-oriented risk-taking activities. Vertically-related equipment suppliers look relatively dynamic in CCC interaction. Even so, the level of flexible-synergy effects is much lower than, and that of market-like effects is as low as, vertically-related parts/material suppliers. Hence, the nature of cooperation that was based on either supply or technology linkages generated a very different way of mixing cooperation-promoting and competition-generating measures for market-linked adjustments. Long-term relations bound by technology were less cooperative, dynamic, and responsive to competition-generating measures. CCC interaction operated more effectively and dynamically between more frequently interacting, unbounded but flexibly cooperating partners, as found in the case of parts and materials procurement.

Another influential factor was the characteristics of partners, which were associated with their positions in the market. Horizontally-related suppliers in both functional areas were as capable as semiconductor companies. But the market conditions clearly differentiated the ways they developed CCC interaction. The parts and material market was more competitive than the oligopolistic equipment market. This difference in market structure differentiated power positions. Facing some competition, parts/material suppliers developed asymmetric power relationships with semiconductor companies, favoring the latter, hence they needed to provide diverse benefits to satisfy the extra needs of semiconductor companies and to respond better to competition-generating measures. The consequence was that horizontally-related parts/material suppliers successfully mixed cooperation-promoting and competition-generating measures.

In contrast, horizontally-related equipment suppliers were in such an oligopolistic position that they were able to stand in a superior bargaining position to semiconductor manufacturers, which even needed to bargain by threatening to alter procurement sources. Such measures weakened a sense of cooperation, when applied to vertically-related suppliers and intrafirm members. Horizontally-related equipment suppliers confined their operations to the joint development of highly sophisticated equipment, were not willing to develop future-oriented risk-taking projects, and were more likely to conduct basic R&D internally. Although technology was the source of long-term relations, without developing a sense of cooperation and future-oriented risk-taking activities, CCC interaction became quite weak.

The most dynamic group performing CCC interaction in both functional areas were the vertically-related suppliers. Given their weaker technological position, they tried to compete against technologically competitive suppliers by providing extra services and coordinating delivery, quality control, and equipment maintenance. Both equipment and parts/material suppliers behaved similarly in generating flexible-synergy effects. They strongly emphasized cooperation and relied on strong human bonding. One crucial difference was that vertically-related parts/material suppliers maintained a better bargaining position than equipment suppliers. These parts/material suppliers and semiconductor companies interacted and mutually influenced each other, and developed future-oriented risk-taking activities. Given the far more sophisticated nature of technology in manufacturing equipment, vertically-related companies could not stand in the same bargaining position as parts/material suppliers. They needed diverse inputs from semiconductor companies and depended on them for maintaining technological capability.

Under the suppressive control of semiconductor companies, they were forced to respond to more competition-generating measures than parts/material suppliers, though they still interacted dynamically with semiconductor manufacturers.

Hence, frequent interaction, unbounded cooperation, less organizationally-restricted relations, and a more competitive market allowed the dynamic development of CCC interaction. Parts/material suppliers relied more strongly on CCC interaction than equipment suppliers. The lack of interaction, restricted cooperation, and organizational restrictions were not quite conducive to developing flexible cooperation, due in great part to the oligopolistic market structure determined by the highly sophisticated nature of technology and the functional needs of semiconductor manufacturing equipment. In the future, however, the oligopolistic market structure may become a less conspicuous reason for competition in the highly technologically sophisticated manufacturing equipment market, and may allow more dynamic development of CCC interaction.

Chapter Five

Effects of Competition-oriented CCC Interaction in Sales and Internalized Interaction in R&D

As reported earlier, over 70% of transactions in semiconductor sales involved competitive-cum-cooperative (CCC) interaction, while over 90% of the R&D budget of semiconductor manufacturers was allocated to intrafirm R&D activities. In contrast, CCC interaction had only a 7.6% share in the total R&D budget. Why do semiconductor companies choose CCC interaction for sales activities, yet choose internalized interaction for R&D activities?

Chapter Three noted that sales activities involve competition-oriented CCC interaction. Severe market competition compels semiconductor companies to develop cooperative relations with customers to obtain new project information earlier than competitors, assist product development and planning, increase the probability of winning contracts, and reduce uncertainty in sales activities. Still, mutual influence and a sense of cooperation and trust play a significant role, developing flexible cooperation between sellers and buyers.

Competition-oriented CCC interaction is different from cooperation-oriented one, but it would still delicately mix cooperation and competition. Such dynamics should be generated by a unique combination of two types of effects: market-like effects derived from the severe market competition semiconductor companies face, and flexible-synergy effects derived from cooperation that they develop to reduce market uncertainty. As was the case in procurement relations, interaction between vertically-related customers and semiconductor manufacturers would be more oriented to achieving flexible-synergy effects than to market-like effects. In contrast, interaction between horizontally-related customers and semiconductor manufacturers would generate more market-like effects.

CCC interaction, however, is in no way part of the dynamics of R&D activities. This chapter will also briefly examine why semiconductor companies choose intrafirm relations to achieve goals in R&D.

To examine flexible-synergy[1] and market-like effects[2], I compared the degrees of benefits and pressures respectively that managers at the semiconductor manufacturers perceived from interacting with and implementing competition-generating measures for each type of partner. Questions were worded differently to suit the functional areas, and new ones were added. This chapter calls on the same methodology as Chapter Three.

A.   Sales

We have noted in Chapter Three that horizontally-related (50.6%) and vertically-related (22.2%) customers engaged in competition-oriented CCC interaction with semiconductor companies, showing a medium+ level of cooperation (see Table 3.7). How do horizontally-related customers keep their relations with semiconductor companies dynamic? In contrast, intrafirm sales

---

[1] Respondents were asked to select the most important company representative for each type of interaction, and to answer the following questions:

(1) What is the degree of benefit that your company enjoys from the indicated company in the following issues:
   (a) Acquiring technological information
      (1. hardly any, 2. a little, 3. some, 4. a good deal, 5. a great deal)
   (b) Acquiring product-market information
      (1. hardly any, 2. a little, 3. some, 4. a good deal, 5. a great deal)
      [Note: On a questionnaire concerning R&D, I asked about acquiring R&D information.]
   (c) Obtaining business, owing to long-term mutual experience
      (1. hardly any, 2. a little, 3. some, 4. a good deal, 5. a great deal)
      [Note: On a questionnaire concerning R&D, I asked about obtaining joint R&D projects.]
   (d) Enhancing trust and cooperation with each other
      (1. hardly any, 2. a little, 3. some, 4. a good deal, 5. a great deal)
   (e) Reducing R&D risk
      (1. hardly any, 2. a little, 3. some, 4. a good deal, 5. a great deal)
      [Note: This question was only asked on a questionnaire concerning R&D.]
   (f) Coordinating delivery and quality control
      (1. hardly any, 2. a little, 3. some, 4. a good deal, 5. a great deal)
      [Note: On a questionnaire concerning R&D, I asked about coordinating R&D.]
   (g) Receiving stable demand
      (1. hardly any, 2. a little, 3. some, 4. a good deal, 5. a great deal)
      [Note: This question was only asked on a questionnaire concerning sales.]
   (h) Providing services that satisfy your company's needs
      (1. hardly any, 2. a little, 3. some, 4. a good deal, 5. a great deal)
      [Note: On a questionnaire concerning R&D, I asked about sharing human resources and equipment for a R&D project.]
   (i) Keeping corporate secrets
      (1. hardly any, 2. a little, 3. some, 4. a good deal, 5. a great deal)
   (j) Stimulating creativity
      (1. hardly any, 2. a little, 3. some, 4. a good deal, 5. a great deal)
   (l) Developing future-oriented risk-taking R&D projects
      (1. hardly any, 2. a little, 3. some, 4. a good deal, 5. a great deal)

from semiconductor divisions to others showed a strong+ level of cooperation and interdependence, whose characteristics were found to be similar to cooperation-oriented vertical interaction in parts and materials. Why does internalized interaction in sales show similar characteristics? Inquiry on flexible-synergy and market-like effects provides insight. Since the severity of market competition – which is so vital – appears in market-like effects, I shall analyze them first.

1.   Market-like Effects

a.   *Pressure from Market-linked Adjustments*

Market-like effects in sales activities reflect the severity of pressure exercised by buyers on semiconductor manufacturers. Questions were asked about the severity of market pressures for selling at stable and unfavorable prices, and keeping a high priority of business with the buyer. These three questions were chosen, assuming that needs for developing cooperative relations did pressure a seller to price as low as possible and continuously maintain the low price. And the seller's attempt to develop pseudo-long-term relations forced him to maintain a high priority of business with the buyer. It prevented a seller to change the business priority of the customer, even when another customer offered a higher price and demanded priority. These were unfavorable conditions for semiconductor manufacturers. As a matter of fact, higher scores of market-like effects in Table 5.1 suggests that market competition was more severe than in the procurement-related functional areas.

---

[2] In a questionnaire concerning sales, respondents were asked to select the most important company representative for each type of interaction, and to answer the following questions:

(1) What is the degree of pressure that your company receives from the indicated company on the following issues? Pressures for:
  (a) Selling at stable prices
    (1. hardly any, 2. a little, 3. some, 4. a good deal, 5. a great deal)
  (b) Selling at unfavorable prices to your company
    (1. hardly any, 2. a little, 3. some, 4. a good deal, 5. a great deal)
  (c) Keeping a high priority of business with the company
    (1. hardly any, 2. a little, 3. some, 4. a good deal, 5. a great deal)
  (d) Implementing contracts flexibly, according to market conditions
    (1. hardly any, 2. a little, 3. some, 4. a good deal, 5. a great deal)
  (e) Maintaining strict standards and a tough relationship
    (1. hardly any, 2. a little, 3. some, 4. a good deal, 5. a great deal)
    (Note: On a questionnaire concering R&D, a question was asked about benefit rather than pressure.)
  (f) What are some of the methods that the indicated company uses to evaluate your com pany? (Please list concretely.)
    (   )

Table 5.1 Market-like Effects by Type of Interaction in Sales

PRESSURE	SPOT INTERACTION (N = 7)	HORIZONTAL INTERACTION (N = 6)	VERTICAL INTERACTION (N = 7)	INTERNALIZED INTERACTION (N = 6)
TYPE OF INTERACTION				
A. From Market-linked Adjustments				
Selling at Stable Prices*	2.86	3.50	3.57	3.17
Selling at Unfavorable Prices*	3.00	3.83	3.50	3.83
Keeping High Priority of Business*	3.00	3.71	3.71	3.80
B. From Resource Interdependence Adjustments				
Implementing Contracts Flexibly*	3.00	3.00	2.86	3.17
C. From Reevaluation Adjustments				
Maintaining Strict Standards and Tough Relationship*	3.00	3.50	2.67	3.00

Note: *1.0 means very weak and 5.0 means very strong. Scores are categorized as follows: very weak (mean value = 1.0 - 1.4); weak (1.5 - 1.9); weak+ (2.0 - 2.4); medium (2.5 - 2.9); medium+ (3.0 - 3.4); strong (3.5 - 3.9); strong+ (4.0 - 4.4); and very strong (4.5 - 5.0).

According to the table, horizontally-related and intrafirm customers put the toughest pressure on keeping prices low (at a strong level), and vertically-related customers a little less (at a strong level). But these low prices were kept low continuously by horizontally- and vertically-related customers (at a strong level), while intrafirm customers were a little more flexible in price changes (at a medium+ level). These three types of partners made sure that they had a high priority in business transactions (at a strong level), preventing semiconductor companies changing customers on the basis of price. Customers in the spot market were the weakest in applying such pressures (at medium to medium+ levels).

A combination of these three indicators for market-linked adjustments provides a very interesting contrast among all types of partners. Horizontally-related customers imposed the toughest conditions for price, while two other types of partners were tough but with some flexibility in price-related conditions. Customers in the spot market were the easiest for semiconductor companies to transact with.

Since horizontally-related customers usually bought a large amount of semiconductors with stable demand and provided opportunities to develop new devices, they achieved the strongest bargaining position vis-à-vis semiconductor companies. Actually, they were the only type of partner in all functional areas, who could stand in a stronger bargaining position than semiconductor companies (see Table 3.7). They forced the latter to keep the priority of business with them and continuously sell semiconductors at unfavorable prices. Despite unfavorable conditions, every semiconductor company struggled to establish cooperative relations with them, only to achieve pseudo-long-term relations which looked like long-term relations, but in reality were but unstable short-term contracts that had been won consecutively.

Semiconductor manufacturers faced severe pressure even from technologically less-sophisticated vertically-related customers. Despite the fact that customers were less attractive product-innovators and bulk purchasers, semiconductor companies were forced to sell their products continuously at quite unfavorable prices even with a high priority of business. The only relief for semiconductor companies was to price a little higher than for horizontally-related customers.

Intrafirm customers pressured semiconductor divisions as much as horizontally-related companies did, except that prices did not have to be kept stable. Since intrafirm customers were members of the same company, semiconductor divisions usually interacted and negotiated more frequently, creating an easier ground for more improvements with sometimes lower and sometimes higher prices whenever necessary.[3]

As I have noted in Chapter Three, between these divisions a quite intensive interaction and a high degree of mutual influence took place, generating a significant degree of flexible-synergy effects. These benefits were attractive for both partners and became incentives for more business relations. But due to severe competition that intrafirm customers faced in their own product markets, they also tended to be quite strict in dealing with semiconductor divisions.

Unlike the common understanding that the market is where the most severe pricing takes place, I found the opposite. Semiconductor manufacturers can price higher with customers in the spot market, choose their customers much more freely, and price flexibly. These findings suggest that CCC and intrafirm interaction can generate a much more severe competitive environment than the spot market.

Keeping a high priority of business with some selected customers reveals the importance of maintaining pseudo-long-term relations. Winning consecutive contracts required semiconductor manufacturers to continuously keep cooperative relations and obtain detailed information long before competitors did. They competed over the degree of cooperation, placing themselves in extremely vulnerable positions vis-á-vis customers. In contrast, the lack of need for developing cooperative relations in the spot market reduced a

---

[3] Interviews with managers in Companies R, T, and V.

sense of competition among manufacturers, since they could always look for some other customers. In sales activities where competition-oriented CCC interaction is dominant, the salespersons' behavior centers on the development of pseudo-long-term relations, and intensifies competition over prices and cooperative behavior as well. Hence, competition-oriented CCC interaction, delicately mixing cooperation and competition, generates more complicated competitive pressure than the spot market.

b.   Pressure from Resource Interdependence and Reevaluation Adjustments

For companies practicing just-in-time production, a short-term contract (averaging about six months) has pragmatic meaning as a measure for adjusting delivery time and quantity, and guaranteeing purchase and delivery. The semiconductor company's compliance to the flexible operation of contracts becomes indispensable in satisfying customer needs. Since just-in-time production is a customary practice in Japan, there is no clear pattern in the flexible application of contracts. Pressure for resource interdependence is similar for all types of partners.

Pressure for reevaluation adjustment is quite severe. Buyers make use of severe competition among semiconductor companies, and demand high standards and tough relationships. Some complaints at the end of the 1980s by Japanese buyers against foreign semiconductor manufacturers, provide insight about the strictness of standards. (1) Failure analysis by foreign companies took twice as long as by Japanese companies, partly because of dependence on foreign laboratories. The worst case was that of a foreign company requiring ten months for analyzing a failure, while the Japanese company took two months. Moreover, the failure rate was not taken as seriously as Japanese companies did. (2) Products from foreign-based suppliers were likely to suddenly reveal defects in an entire lot. (3) Foreign companies changed specifications without consultation or even notice, while Japanese companies always consulted or informed. (4) Although foreign companies were quite creative, they took a longer time to develop products (INSEC, 1991:3). (5) Japanese and foreign buyers also had a different perception about reasonable delivery time from the date of order. Out of 46 Japanese buyers and 23 foreign semiconductor manufacturers, 12% and 4% respectively responded "less than two weeks;" 60% and 4%, "two weeks to one month;" 28% and 50%, "one to two months;" and 0% and 34%, "two to four months." Japanese buyers expected to have products delivered much more quickly than foreign semiconductor manufacturers anticipated. (6) Similarly, acceptable time for delay in delivery differed greatly. Out of 46 Japanese users and 23 foreign semiconductor manufacturers, 15% and 0% respectively responded "less than a day;" 43% and 9%, "less than two days;" 34% and 17%, "within a week;" 6% and 44%, "within two weeks;" and 2% and 30%, "within a month" (INSEC, 1988:8). Japanese buyers tolerated less time for delivery delay than foreign semiconductor manufacturers perceived. Evidence clearly indicated that Japanese buyers set much stricter standards than foreign semiconductor manufacturers.

Customer evaluation of a semiconductor manufacturer is quite important. Customers' satisfaction definitely weighs heavy. They often inform the seller about evaluation results and sometimes issue a warning, expecting improvement in performance. For example, one horizontally-related customer evaluated semiconductor companies by the grades, A, B, C, D, and F in each specified area. A meant a status equal to a group member with free access to training and research facilities. B meant a good performer requiring further efforts to be treated as a group member. C meant a standard performer whose position might be threatened with the appearance of a better performer. D meant a warning for improving performance. Failure to improve within a fixed period would result in F, termination of the relationship. Another customer gave a simple ranking to fifteen companies such as "tenth in delivery accuracy" or "eleventh in failure rate." If improvement did not materialize, the next contract would go to a higher-ranked or new company. Once a relationship was terminated, a semiconductor company had to convince the customer all over again that it could do much better than before and better than competitors.

Another unusual system of evaluation is site inspection. Customers can inspect the production processes of semiconductor manufacturers. In 1980, 60% of 67 IC producers and 100% of 67 bipolar producers that responded to a survey accepted customer's inspection of production facilities (Kikai Shinko Kyokai Keizai Kenkyusho, et al., 1980:72). This practice came to be accepted as a legitimate evaluation method in an earlier period of development of the Japanese electronics industry, namely, when Nippon Telegraph and Telephone Public Corporation was the major purchaser of telecommunication devices. The Japanese government's industrial policy was encouraging the development of new products by providing R&D funding and enforcing high quality standards. Products were purchased only from the companies that met these high standards, and as a way of ensuring high quality, production facilities were inspected. Inspecting production processes had become a legitimate method of customer evaluation (Aida, 1992:403).

According to Table 5.1, horizontally-related customers were pressuring semiconductor companies for much stricter standards and tougher relations than any other type of partner (at a strong level), while both intrafirm and spot-market customers exercised somewhat less pressure (at a medium+ level). Vertically-related customers were not forcing such strict standards and tough relationships (at a medium level).

Bulk purchases, forefront technologies, and the performance-oriented logic of continuity, all placed horizontally-related customers in a strong position, enabling them to enforce strict standards and tough relations. Another type of customer that exercised tough relations were intrafirm customers who often procured high-risk integrated circuits (see Common Goals and Interests in Table 5.2). Since success in developing a new product determines engineers' future promotion, high standards were set. Furthermore, due to frequent interaction, their demands tended to be much more severe than the demands by outside companies.[4] For example, a researcher in a semiconductor division developed a CMOS LSI, but faced intrafirm customers' unwillingness to use the device. Since the product had been developed to comply with their needs, the response was unexpected. Each division was a profit center, acting in light of its own interests, and did not accept what it perceived to be a product of inferior quality. Consequently, the researcher approached an outside company willing to test the device by incorporating it into a new calculator, which became a best seller because of its low battery requirement (Nakagawa, 1989:83). Usually, in the same company other divisions adopt the best device available and do not give preference to the semiconductor division, unless some emergency situation determines internal procurement.[5] Hence, the profit center orientation of each division creates a market-like environment.

The unusually tougher relations imposed by customers in the spot market might be due to severe competition over standardized products. Surprisingly, there appeared to be less stringent toughness applied by vertically-related customers, owing partly to their emphasis on trust, cooperation and interdependence.[6] Less technologically sophisticated, but important, customers with human-relations-oriented logic of continuity were not quite successful in implementing tough relations.

Thus, when the development of pseudo-long-term relationships was important to win competition, semiconductor manufacturers competed severely

---

[4] Interviews with managers in Companies Q, R, and T.

[5] Interviews with managers in Companies Q, R and T.

[6] Interview with Company Q manager.

in facing customers' tough demands. They were forced to sell continuously at quite low and stable prices with strict standards, and were confronted with tough negotiations and relations. Pressures were mainly generated from market-linked and reevaluation adjustments.

Owing to the highly attractive nature of their orders, horizontally-related

Table 5.2 Flexible-synergy Effects by Type of Interaction in Sales

BENEFITS	TYPE OF INTERACTION			
	SPOT INTERACTION (N = 7)	HORIZONTAL INTERACTION (N = 7)	VERTICAL INTERACTION (N = 7)	INTERNALIZED INTERACTION (N = 6)
A. From Long-term Mutual Accumulation				
Acquiring Technological Information*	1.86	3.29	3.71	3.50
Acquiring Product-market Information*	2.43	3.14	3.57	3.83
Obtaining Business, Owing to Long-term Mutual Experience*	2.71	3.43	3.00	3.17
Enhancing Trust and Cooperation*	2.14	3.29	3.43	3.07
B. From Satisfying Mutual Expectations				
Coordinating Delivery and Quality Control*	2.14	3.00	2.71	3.33
Receiving Stable Demand*	2.43	3.29	3.00	3.33
Providing Services to Satisfy Customer Needs*	2.29	3.00	3.29	3.33
Keeping Corporate Secrets*	2.00	2.83	3.17	3.50
C. From Future-oriented Risk-taking Activities				
Stimulating Creativity	2.00	3.57	3.14	3.17
Developing Future-oriented Risk-taking Projects	2.33	3.17	3.43	3.17
D. Common Goals and Interests				

	- Cost
	- Competition
	- Extra earnings
		- Cost
		- Competition
		- Stable supply
		- Mass-
		   production
		- Quality
		- Delivery
		- Service
		- Information
		- Product market
		   share
		- Mutual profit
			- Cost
			- Competition
			- Stable supply
			- Mass-
			   production
			- Quality
			- Delivery
			- Service
			- Information
			- Product market
			   share
			- Mutual profit
				- Cost
				- Competition
				- Stable supply
				- Mass-
				   production
				- Quality
				- Delivery
				- Service
				- Information
				- Product market
				   share
				- Mutual profit
				- High risk ICs

Note: * 1.0 means very weak and 5.0 means very strong. Scores are interpreted using the following categories: very weak (mean value = 1.0 - 1.4); weak (1.5 - 1.9); weak+ (2.0 - 2.4); medium (2.5 - 2.9); medium+ (3.0 - 3.4); strong (3.5 - 3.9); strong+ (4.0 - 4.4); and very strong (4.5 - 5.0).

customers made the best use of competition-generating measures, and clearly succeeded in setting the severest pressures on prices and standards. Intrafirm customers forced semiconductor divisions with high priority of business to price as low as horizontally-related customers did. This was because the profit-center orientation of each division created a market-like environment. But being a member of the same company, prices could fluctuate and standards tended to be slightly less strict than those of horizontally-related companies. Vertically-related customers forced semiconductor companies to maintain a low price and treated them with high priority of business, but prices could be somewhat higher than for horizontally-related companies. Their weak techno-logical capability, the unattractiveness of their orders to semiconductor com-panies, as well as their human-relations-oriented approach reduced their capa-bility to implement strict standards and tough relations. Even customers in the spot market for standardized semiconductors were as strict and tough as intrafirm customers, but the spot market became an easy place for semiconductor com-panies to charge higher prices by switching transaction partners. The mecha-nisms for generating low cost and high quality supply are rather better devel-oped in CCC interaction than in the spot market.

2.   Flexible-synergy Effects

a.   Benefits from Long-term Mutual Accumulation

Since the semiconductor industry requires heavy initial investments and the product market fluctuates frequently, manufacturers try their best to stabilize demand for their products, so that they can properly plan investment and R&D expenditures. The best strategy for semiconductor manufacturers to survive even in a down-turn of demand is to remain one of the best suppliers of their customers. Semiconductor managers identified the following condi-tions: (1) to obtain trust by satisfying customers' demands as much as pos-sible; (2) to maintain technological leadership and provide the best products; (3) to reduce the time requirement for new product development; (4) to shorten the required time for production; (5) to deliver products punctually; (6) to have an extra production capability so that increased demand can be swiftly accommodated; (7) to ensure the high quality and reliability of their prod-ucts; (8) to perform problem analysis fast and provide technical support; and (9) to keep improving beyond the targets set by competitors (Sakamoto, 1990).[7]

---

[7] Interviews with managers in Companies W and Q.

Since severe competition over a custom-made semiconductor takes place at the designing stage, aggressive contacts and sales activities continue until a customer chooses one company and orders an engineering sample at a certain price. Although the cost of developing the sample is shouldered by the customer, the "black box" where both buyers and sellers interact to develop detailed specifications, requires intensive exchange of information previous to producing the sample. The more effective the interaction, the more certainly a contract will follow. Once the sample is found acceptable, the manufacturer is ready for mass production.[8]

Even in the mass production stage, cooperation between the customer and the semiconductor manufacturer remains intensive, since within one year of the production contract the customer will make diverse demands, increasing or reducing production, changing specifications, and sometimes requesting more services. The ways in which semiconductor companies handle these demands greatly influence future business transactions. Experience accumulated during the first contract becomes the base for the next contract. After two or three consecutive contracts, the buyer will decide whether the manufacturer is sincere, capable, trustworthy, flexible, and cooperative. Needless to say, the manufacturer on its part has to perform best in quality, price, delivery and services. To do so over and over again requires intensive communications.

The more experience the semiconductor manufacturer accumulates in dealing with the customer, the more it succeeds in being trusted, making transactions easier and increasing the probability to win additional contracts. According to Table 5.2, semiconductor companies felt that they generated more business by developing pseudo-long-term relations with horizontally-related customers than any other type of partner (at a medium+ level). With the same customers, they also experienced that trust and cooperation were enhanced through continued relations (at a medium+ level). Accumulating experience in repeated contracts and nurturing trust and cooperation were found effective with horizontally-related customers.

In contrast, pseudo-long-term relations with vertically-related customers resulted in enhancing trust and cooperation more effectively than any other type of partner (at a medium+ level). It suggests that good relations with them could be more easily achieved by adopting the human-relations-oriented logic of continuity rather than the performance-oriented strategy. Intrafirm customers' responses were between these two extremes, while spot-

---

[8] Interview with Company Q manager.

market customers were rather poor in generating such benefits (at weak+ to medium levels).

Vertically-related customers facilitated the flow of technological information best (at a strong level), while intrafirm customers helped the semiconductor division to acquire product-market information (at a strong level). Although the questions addressed to semiconductor companies were about benefits in acquiring product-market and technological information, the answers also suggested benefits in acquiring information on projects. For horizontally-related customers, project information was a crucial bargaining edge; control over this information enabled them to stimulate competition among semiconductor companies. Hence, they did not easily give away this advantage, scoring quite low in the indicators of Acquiring Technological Information and Acquiring Product-market Information (both at a medium+ level) relative to vertically-related and intrafirm customers (both at a strong level).

b.   *Benefits from Satisfying Mutual Expectations*

When market competition is very severe, satisfying customers' demand becomes most important but difficult. Semiconductor companies must perform well in coordinating delivery and quality control, providing services to satisfy customer needs, and keeping corporate secrets. However, high performance in these practices with horizontally-related customers did not necessarily turn into benefits (at a medium+ level), since every competitor performed equally well in satisfying mutual expectations.

Semiconductor companies, however, reaped the benefit of having a little more stable demand from horizontally-related customers than other types of customers (at a medium+ level). Vertically-related customers, who were less bulk purchasers, could not provide such benefits to semiconductor companies, but they responded a little better to extra services offered by semiconductor manufacturers (at a medium+ level).

An unusual case is intrafirm sales. Interaction between the semiconductor and other divisions of the same company showed quite complex characteristics. On the one hand, other divisions acting as customers demanded strict standards and a tough relation over prices. On the other hand, the semiconductor division and its intrafirm customers were the most cooperative in coordinating delivery and quality control, enjoying stable demand, providing services to satisfy customer needs, and keeping corporate secrets (all at medium+ to strong levels). According to Table 3.7, they developed a strong+ level of cooperation and interdependence. Other than keeping corporate secrets, it is quite unusual to find so many flexible-synergy effects generated by internalized interaction. Intrafirm sales seem to show characteristics different from the generally understood relations in hierarchical or competition-oriented CCC governance. The characteristics seem to be rather closer to cooperation-oriented vertical relations. The organizational structure appears to function as a buffer against the severe competition taking place in the market and provides a favorable environment for cooperation. Customers in the spot market were not much interested in benefiting from flexible-synergy effects.

c.   Benefits from Satisfying Future-oriented Risk-taking Activities

Building a base for future cooperation is an indispensable part of maintaining pseudo-long-term relations. Semiconductor companies have to gather information about future projects and keep making new and interesting proposals to customers. The more easily semiconductor companies obtain stimulation for new ideas from customers, the better able they are to propose new and interesting semiconductors and the more business opportunities are created.

The situation, however, was not that simple. Horizontally-related customers, as equally future-oriented as intrafirm customers, stimulated creativity the most (at a strong level), but they were somewhat less willing to cooperate in future-oriented risk-taking activities (at a medium+ level) than vertically-related customers. They committed to future-building at an intermediate level, because they operated more independently, given their technological capability. In contrast, less-technologically-sophisticated vertically-related customers needed help from semiconductor companies in developing innovative products, and were most willing to cooperate in future-oriented risk-taking activities (at a medium+ level). However, they were not as stimulating partners for creativity (at a medium+ level) as horizontally-related customers. In addition to weak technological and managerial capability, there was another reason why vertically-related customers deliberately adhered to the human-relations-oriented logic of continuity. For them, harmonious human relations kept open the channels of business and technological ideas, and thus facilitated business opportunities for semiconductor companies.

The less attractive position of intrafirm customers in stimulating creativity and risk-taking for future-oriented activities (at a medium+ level) resulted from the rather small proportion of high-risk IC business relative to other application-specific integrated circuits. Nonetheless, they also engaged in future-oriented risk-taking activities at an intermediate level, similar to horizontally-related customers. Again, companies operating in the spot market provided the least creative stimulation and the least efforts for future-oriented risk-taking activities (both at a weak+ level).

3. Summary: Tough Horizontally-related and Highly Cooperative Intrafirm Customers

Semiconductor manufacturers wish to establish long-term relations with companies that make bulk purchases based on stable demands, since they allow manufacturers to plan investment and budget allocations for R&D. The severity of competition among semiconductor companies, however, makes long-term relationships difficult to nurture. Facing competition, they try to win contracts consecutively and at best develop pseudo-long-term relations. The pattern of such development differs with each partner.

Table 5.3 shows the performances of each type of interaction in generating flexible-synergy and market-like effects. By comparing the category of low performance, all types of partner other than customers in the spot market achieved a similar level, with only a slightly better performance by intrafirm customers.

Comparing indicators in the category of top performance, horizontally-related customers implemented relatively heavy competition-generating measures, suggesting that they were taking advantage of market competition the most. Their capability to develop new products and make bulk and stable purchases over a longer period made them the most attractive customers. But simultaneously, due to severe competition, they became the toughest partners from whom to win a contract. Especially, their toughness in prices, standards and relations had distinct characteristics. Strong pressure for lower price was not the characteristic of only horizontally-related customers, but these customers were the only type of partner who marked all market-linked adjustment measures - stable low prices and high priority given to their business - as the top priority for semiconductor manufacturers to meet. The low benefits of semiconductor companies in acquiring technological and product-market information suggested difficulty in obtaining the project information that functioned as a bargaining edge to solicit more cooperative behavior from the manufacturers. The low benefits in satisfying customer needs also suggested the difficulty of satisfying demands simply by operational performance. Nonetheless, horizontally-related customers emphasized the importance of cooperation, expected that continued relations would strengthen a sense of cooperation and trust, and facilitated transactions. They also en-

Table 5.3 Performance by Type of Interaction in Sales

		TYPE OF INTERACTION		
PERFORMANCE	SPOT INTERACTION	HORIZONTAL INTERACTION	VERTICAL INTERACTION	INTERNALIZED INTERACTION
Top Performance (Top in Ranking)		Obtaining business, owing to long-term mutual experience Enhancing trust and cooperation Receiving stable demand" Stimulating creativity *Selling at stable prices** *Selling at unfavorable prices"* *Maintaining strict standards and tough relationship*	Acquiring technological information Enhancing trust and cooperation Providing services to satisfy customer needs Developing future-oriented risk-taking projects *Selling at stable prices**	Acquiring product-market information Coordinating delivery and quality control Receiving stable demand% Providing services to satisfy customer needs Keeping corporate secrets *Selling at unfavorable prices"*
Low Performance (Third and Fourth in Ranking)	Acquiring technological information Acquiring product-market information Obtaining business, owing to long-term mutual experience Enhancing trust and cooperation Coordinating delivery and quality control Receiving stable demand Providing services to satisfy customer needs Keeping corporate secrets Stimulating creativity Developing future-oriented risk-taking projects *Selling at stable prices* *Selling at unfavorable prices* *Keeping high priority of business*	Acquiring technological information Acquiring product-market information Providing services to satisfy customer needs Keeping corporate secrets	Coordinating delivery and quality control Receiving stable demand *Selling at stable prices* *Maintaining strict standards and tough relationship*	Enhancing trust and cooperation *Selling at stable prices*

Note: This table is based on mean values in Tables 5.1 and 5.2.
Regular letters indicate flexible-synergy effects; *italic letters* indicate market-like effects.
* The top position is shared between vertical and horizontal interaction.
% The top position is shared between vertical and internalized interaction.
" The top position is shared between horizontal and internalized interaction.

gaged in future-oriented risk-taking activities at an intermediate level, though not as strongly as vertically-related customers did.

A complex combination of both market-like and flexible-synergy effects, especially based on the performance-oriented logic of continuity, generated efficiency on the part of horizontally-related customers. Winning a contract by meeting these tough measures drastically increased the probability of winning the next contract, but a few unexpected incidents or the appearance of a better performing competitor would readily terminate the relationship. Hence, interaction between semiconductor manufacturers and horizontally-related customers showed the characteristics of pseudo-long-term relations based on the performance-oriented logic of continuity. They were the core actors in the competition-oriented CCC governance, which coordinated interactive behavior and made cooperation a means of reducing uncertainty and threats, and surviving competition.

Vertically-related customers, minor actors under this governance, did not manifest the distinctive characteristics in competition-generating measures observed in horizontally-related customers. They pressured semiconductor manufacturers on prices, but a little less than horizontally-related customers did. However, the need for obtaining valuable technological and business ideas from semiconductor manufacturers reduced their bargaining edge, while those manufacturers needing to secure future contracts kept close human relations and continuously suggested new semiconductor-demand-creating ideas. Vertically-related customers were very eager to develop future-oriented risk-taking projects jointly. They strongly emphasized the importance of cooperation and trust, based on the human-relations oriented logic of continuity, which also allowed semiconductor companies to reduce the tough pressures exercised by customers. Hence, semiconductor manufacturers cleverly differentiated their strategies between horizontally- and vertically-related customers in order to reduce uncertainty originating in market competition.

Intrafirm customer relations manifested complex characteristics of cooperation and tough relationship. Partners emphasized at times a sense of cooperation as members of the same company and at other times stringent and tough conditions as customers. Two explanations seem to be possible. One is that the semiconductor division was engaged in the business of developing high-risk integrated circuits ordered by other divisions. Intense cooperation was required, while tough relationships allowed it to maintain high level standards. The other is the organizational structure of the company. In many companies each division was operated as a profit center and interacted with other divisions as if they were independent companies; a market-like environment was created inside the company. Similarly to vertical interaction under cooperation-oriented CCC governance, two separate divisions promoted cooperation to generate flexible-synergy effects, while they implemented competition-generating measures to make cooperation flexible and dynamic. Since the proportion of business allocated to high-risk integrated circuits was not big, the latter explanation may be more plausible.

Semiconductor and other divisions made good use of a highly cooperative environment within the organization. But in areas such as pricing and standards, where organizations often faced difficulty to implement strict measures, other divisions introduced tough measures cold-heartedly based on the performance-oriented logic of continuity. Hence, internalized interaction exhibited a complex interplay between competition and cooperation. It was still harmonized under hierarchical governance, but with some effective introduction of a market-like environment.

Customers operating in the spot market were the least capable of pressuring semiconductor manufacturers and failed to purchase at favorable prices. Their incapability to develop pseudo-long-term relations, obtain information in a cooperative manner, and generate flexible-synergy effects, jeopardized opportunities for winning contracts. Hence, when institutional arrangements were made in a way to generate efficiency from a complex mixture of competition and cooperation, but in a pseudo-long-term manner, incapability to do so made companies in the spot market the least effective and attractive partners.

In sales activities, CCC interaction that delicately mixed flexible-synergy and market-like effects was dominant. Its best example was interaction between horizontally-related customers and semiconductor companies. The former made full use of market competition to enhance their gains, while the latter used cooperation as the means for winning contracts and reducing uncertainty in the highly competitive market. Vertically-related customers also engaged in CCC interaction, but their inferior technological and managerial capability and the human-relations-oriented logic of continuity weakened their bargaining position. Introducing a market-like environment, while making full use of the strong cooperative environment within a company, intrafirm customers developed a dynamic interaction with semiconductor divisions. They succeeded in mixing both cooperation-promoting and competition-generating measures even under hierarchical governance. Hence, regarding sales, competition-oriented CCC governance and hierarchical governance with the introduction of a market-like environment were found to operate dynamically.

B.   Dynamics of Intrafirm Relations in R&D

An intrafirm environment is the most effective for developing R&D coop-eration; it generates high levels of flexible-synergy effects and keeps tech-nology secrecy. According to Table 5.4, semiconductor divisions and intrafirm research-related units accumulated mutual experience, strengthened their sense of trust and cooperation, exchanged technological and R&D informa-tion, coordinated their activities smoothly, shared human resources and equip-ment, stimulated innovation, and developed jointly future-oriented risk-tak-ing projects (all at strong+ to very strong levels in Appendix Table 5.1). Good planning, strict standards, tough relations, and frequent evaluations of per-formance introduced very high expectations of achievement (at strong+ level). Such tight control and coordination with a strong sense of cooperation could be realized only inside the corporate organization. Internalized interaction in R&D clearly showed the characteristics of hierarchical governance.

Semiconductor manufacturers developed joint R&D projects with verti-cally-related companies to maintain the technological capability of the latter, often providing them with diverse financial, personnel, and other help, while helping themselves in reducing R&D risk and sharing human resources and equipment (at strong+ level). Vertically-related companies were quite de-pendent, and faced suppressive control by semiconductor manufacturers (see Table 3.7), suggesting that interaction between them operated under an envi-ronment similar to hierarchical governance.

Although horizontally-related companies were eager to develop future-oriented risk-taking projects (at a strong level) and keep tight control over corporate secrets (at a medium+ level), their capability to generate flexible-synergy effects was far less than what was found in intrafirm R&D-related units and even vertically-related companies. But their technological superi-ority led to more R&D cooperation, and operated under cooperation-oriented CCC governance.

Semiconductor companies interacted with companies in the spot market in terms of quick technological acquisition, and did not engage in joint R&D projects. Arm's length transactions were clearly the characteristic of market governance.

C.   Comparison of CCC Interaction in the Procurement of Parts/Materials and Sales

In competition-oriented CCC interaction, market competition is the base of interfirm relations, while cooperation is used as a means of reducing uncer-

Table 5.4 Performance by Type of Interaction in R&D

| | TYPE OF INTERACTION | | | |
PERFORMANCE	SPOT INTERACTION	HORIZONTAL INTERACTION	VERTICAL INTERACTION	INTERNALIZED INTERACTION
Top Performance (Top in Ranking)			Acquiring technological information$^{\%}$ Reducing R&D risk Sharing human resources and equipment$^{\%}$	Acquiring technological information$^{\%}$ Acquiring R&D information Obtaining joint R&D projects, owing to long-term mutual experience Coordinating R&D Sharing human resources and equipment$^{\%}$ Keeping corporate secrets Stimulating creativity Developing future-oriented risk-taking projects *Maintaining strict standards and tough relationship*
Low Performance (Third and Fourth in Ranking)	Acquiring technological information Acquiring R&D information Obtaining joint R&D projects, owing to long-term mutual experience Reducing R&D risk Coordinating R&D Sharing human resources and equipment Keeping corporate secrets Stimulating creativity Developing future-oriented risk-taking projects *Maintaining strict standards and tough relationship*	Acquiring technological information Acquiring R&D information Obtaining joint R&D projects, owing to long-term mutual experience Coordinating R&D Sharing human resources and equipment	Keeping corporate secrets	

Note: This table is based on mean values in Appendix Table 5.1.
Regular letters indicate flexible-synergy effects; *italic letters* indicate market-like effects.
$^{\%}$ The top position is shared between vertical and internalized interaction.

tainty in market competition. In cooperation-oriented CCC interaction, co-operation is the base of interfirm relations, while competition is used as a means of generating flexibility in cooperation. In some performances, irrespective of these differences in approaches, both types of interaction achieve the same result, although the manner in which they achieve this result differs by the type of partner in each functional area. But in some performances, given mechanism differences between cooperation- and competition-oriented CCC interaction, particular types of partner are more capable than others to achieve higher performance.

Table 5.5 compares CCC benefits between the procurement of parts/materials and sales. Irrespective of differences in implementing cooperation-promoting and competition-generating measures, CCC interaction was quite effective in suppressing prices. In case of parts/materials, both vertically- and horizontally-related suppliers were pressured the most by semiconductor companies to price low. Similarly, horizontally-related customers in sales, offering highly attractive orders to semiconductor manufacturers, made use of severe market competition to enhance cooperation and to keep prices steadily low. In contrast, vertically-related customers, due to their technological weakness and the less attractive nature of their orders, could not bargain as effectively as horizontally-related customers. Nonetheless, both types of customers in negotiating prices were making full use of severe market competition. In the case of sales, severe market competition was the direct reason for keeping prices low, but competition among semiconductor companies to develop a sense of trust and cooperation and establish pseudo-long-term relations contributed to reducing prices much more so. This means that cooperation also had some relevance in generating low prices.

In short, CCC interaction harbored a mechanism to implement much lower prices than the spot market. Cooperation-oriented CCC interaction made an effective use of competition-generating measures to reduce prices, but the application of these measures was not possible without cooperation. In contrast, competition-oriented CCC interaction effectively mixed both severe market competition and the seller-initiated cooperative behavior for developing pseudo-long-term relations, and succeeded in achieving low prices.

Different types of partners accomplished such a result by making use of a different mix of cooperation-promoting and competition-generating measures. For horizontally-related suppliers in parts and materials procurement, the result was largely achieved by implementing strategic pricing which involved cooperation in planned price reduction, while for vertically-related suppliers, a sense of cooperation and trust in long-term relations had a deep influence. Given the different characteristics of partners, semiconductor compa-

Table 5.5 Comparison of CCC Benefits by Type of Interaction in Procurement of Parts and Materials and in Sales

BENEFITS	PROCUREMENT OF PARTS AND MATERIALS			SALES		
	FROM VERTICAL AND HORIZONTAL INTERACTION	FROM VERTICAL INTERACTION	FROM HORIZONTAL INTERACTION	FROM VERTICAL AND HORIZONTAL INTERACTION	FROM VERTICAL INTERACTION	FROM HORIZONTAL INTERACTION
A. Flexible-synergy Effects						
1. Benefits from Long-term Mutual Accumulation	Acquiring technological information	Obtaining business, owing to long-term mutual experience		Enhancing trust and cooperation	Acquiring technological information	Obtaining business, owing to long-term mutual experience
2. Benefits from Satisfying Mutual Expectations	Providing stable supplies Receiving services to satisfy needs	Coordinating delivery and quality control^x			Providing services to satisfy customer needs^x	Receiving stable demand"
3. Benefits from Future-oriented Risk-taking Activities	Stimulating creativity Developing future-oriented risk-taking projects				Developing future-oriented risk-taking projects	Stimulating creativity
B. Market-like Effects						
1. Benefits from Market-linked Adjustments	*Purchasing inexpensively**		*Implementing strategic pricing*	*Selling at Stable Price*		*Selling at unfavorable price*
2. Benefits from Resource Interdependence Adjustments	*None*			*None*		
3. Benefits from Reevaluation Adjustments			*Maintaining strict standards and tough relationship*			*Maintaining strict standards and tough relationship*

Note: Regular letters indicate flexible-synergy effects; *italic letters* indicate market-like effects.
^x The top position is shared between vertical and internalized interaction.
" The top position is shared between horizontal and internalized interaction.

nies used different approaches, but cooperation remained the basis for successfully implementing such competition-generating measures.

In sales, since cooperation was the means for reducing uncertainty in unstable relations, it operated differently. Semiconductor companies developed cooperation with customers to obtain information as quickly as possible, offered new ideas, and engaged in joint design of semiconductors. Cooperation became an important means for establishing pseudo-long-term relations. This pattern was especially strong with horizontally-related customers. To the contrary, such effort was considered less important by vertically-related customers. Semiconductor companies interacted with them more on the basis of human-relations oriented logic of continuity, providing better services, keeping closer human relations, suggesting new ideas, and developing future-oriented risk-taking projects. Consequently, semiconductor companies succeeded in securing sales. Vertically-related customers, weak in technological capability, solicited such offers from semiconductor companies, and obtained ideas for their future business.

In both parts/materials and sales, horizontally-related and semiconductor companies interacted more on the basis of the performance-oriented logic of continuity with more stringent application of competition-generating measures, while vertically-related and semiconductor companies built their relations on the human-relations-oriented logic of continuity and cooperation-promoting measures. Irrespective of competition- or cooperation-oriented CCC interaction, differences in the logic of continuity influenced greatly the way to mix cooperation and competition.

This difference in approach also influenced performance outcome. In cooperation-oriented CCC interaction, cooperative gain was much higher and diverse than in the competition-oriented one. Semiconductor companies benefited from stable supplies and extra services of both vertically- and horizontally-related suppliers, and from joint future-oriented risk-taking activities. They also benefited from using both types of partners as the core of information networks, an advantage that could not have been achieved without long-term mutual accumulation of experience. Both types of suppliers fulfilled very well the logic of continuity. But vertically-related suppliers provided much stronger flexible-synergy effects, while they also priced as low as horizontally-related suppliers, and had a stronger sense that long-term mutual experience was leading to further development of business relations. Consequently, they became more favorable partners for semiconductor companies than horizontally-related suppliers, owing mainly to differences in the capability of generating flexible-synergy effects. This is simply because the vertically-related suppliers' emphasis on cooperation, trust and human-relations-oriented logic of continuity worked more effectively under cooperation-oriented CCC interaction.

In sales, semiconductor companies approached vertically-related customers similarly. Even in competition-oriented CCC interaction, when the human-relations-oriented logic of continuity satisfied the benefits of long-term mutual experience, mutual expectations, and risk-taking activities, interaction became more stable, continuous, and closer to cooperation-oriented CCC interaction. Such an approach by semiconductor companies aimed primarily at stabilizing sales, not at generating dynamic transactional interaction, since human-relations-oriented logic of continuity worked against the mechanism of competition-oriented CCC governance and weakened interactive dynamics.

In contrast, horizontally-related customers in sales were the core of dynamic interaction under competition-oriented CCC governance. The interaction was dynamic because horizontally-related companies relied on the performance-oriented logic of continuity, made the best use of severe market competition, and effectively used seller-initiated cooperation as a road to higher performance. Interaction between semiconductor companies and horizontally-related customers matched well with the mechanism of competition-oriented CCC governance, and consequently enhanced interactive dynamics.

Hence, the data show that the strength of CCC interaction was in its mechanism to control prices by mixing cooperation and competition. However, this outcome could not be achieved unless semiconductor companies acted selectively, taking into consideration market conditions, characteristics of partners, and their technological capability. In the case of sales, horizontally-related customers interacted most dynamically because their behavior fitted competition-oriented CCC governance, while in the case of parts and materials, vertically-related suppliers interacted most dynamically because their behavior fitted best with cooperation-oriented CCC governance.

Chapter Six

CCC Dynamics and Structure of Intra- and Interfirm Relations

The Japanese semiconductor industry has shown a history of remarkable development from its dependence on U.S. technology in the 1950s and 1960s to the world leading position in the 1980s, especially in the area of dynamic random access memory. An important part of the explanation for such dynamic development is cooperative, flexible, long-term interfirm relations. Competitive-cum-cooperative (CCC) interfirm relations came to be developed because of the pressure to cope with turbulent market conditions, while top-level company managers were restricted in their strategic choices by Japan's institutional inheritance of interpersonalism. Such difficulty was overcome not by introducing completely new practices, but by making good use of this inheritance and incorporating goal- and future-oriented elements in the values and norms sustaining interfirm relations. Path-dependent institutional development that gradually took place from the late 1960s through the oil crises was an important source of dynamic development in the Japanese semiconductor industry. Though the industry's golden age in terms of a world market share has ended, it still continues to demonstrate world leadership. The CCC interfirm relations remain a vital source of dynamic development; they determine how each type of partner contributes to the operations of a semiconductor company and consequently the overall structure of intra- and interfirm relations in the industry.

A. Dynamic Mechanism of Competitive-*cum*-cooperative Interfirm Relations

Long-term relations are still a very important source of industrial dynamics, because they mix cooperation and competition, which are, in essence, contradictory. With cooperating partners, semiconductor companies not only intensify cooperation, but also introduce diverse competition-generating measures, implement monitoring and evaluation, and create a market-like environment among selected companies. Competition-generating measures, such as multiple sourcing, strategic pricing, altering procurement sources flexibly, strict standards for evaluation, etc., introduce a sense of uncertainty into cooperative relations, with even the threat of possible termination. These challenges make cooperation highly flexible. Better performers in competition-generating measures are also rewarded with a greater degree of cooperation. CCC interaction based on this delicate mixture of cooperation-promoting and competition-generating measures, inducing flexible-synergy and market-like effects respectively, is made possible by partners engaging in cooperative bargaining to maximize mutual gain.

CCC interaction can be cooperation- or competition-oriented. Cooperation-oriented CCC interfirm relations are predominantly found in the functional areas of procurement, while competition-oriented CCC interfirm relations are in the area of sales. In cooperation-oriented CCC relations, semiconductor companies use competition-generating measures only to the extent that cooperation with partner companies is not jeopardized, so that cooperation remains flexible and dynamic. The maintenance of cooperation becomes the goal of interfirm relations, while competition becomes a means for generating flexibility in cooperative relations. Such a pattern of interaction is extensively found in relationships between semiconductor companies and parts and material suppliers. Equipment suppliers also show this tendency, but due to less frequent nature of purchases, such a pattern is less conspicuous.

In competition-oriented CCC interfirm relations, sellers develop cooperative relations with buyers as a means of winning a contract and reducing sellers' uncertainty in the market. In sales, especially of application-specific integrated circuits, acquiring information about new products, proposing new ideas and suggestions, and winning contracts consecutively with the same buyers are necessary. Accumulated experience and an enhanced sense of trust and cooperation become meaningful additional comparative advantages to technological capability and new ideas. Sellers try to secure their business by developing pseudo-long-term relations, and reduce uncertainty in their sales activities. Pseudo-long-term relations are in reality nothing but unstable short-term contracts that sellers win consecutively, looking as if long-term relations exist. They allow buyers to generate among sellers competition over cooperative efforts and to benefit from seller-initiated cooperative behavior. How successfully a seller develops cooperative behavior becomes one of the important determinants of buyers' selection. In this type of seller-buyer relations, buyers delicately mix cooperation-promoting measures and market-like effects generated by severe market competition.

In theory, semiconductor manufacturers have the freedom to choose any type of partner and determine how they are going to interact. They could transact with companies in the spot market, based on the way the market coordinates diverse actors. Their interaction is one at a time, and does not involve continued interaction. They could use other departments and laboratories in the same company, where hierarchical relations harmonize actors inside the organization. Its distinct characteristic is the use of bureaucratic control and coordination when relations between major and minor supporting units are involved, and the promotion of strong cooperation and mutual influence when two major units interact, especially in R&D. Yet, for most functional areas, semiconductor manufacturers transact with long-term and pseudo-long-term partners and develop CCC interfirm relations. These actors are harmonized under CCC interfirm governance mixing cooperation and competition in interactive behavior.

The Japanese institutional tradition is characterized by interpersonalism, where each individual understands the expectations of the other and adjusts himself in accordance with his/her expectations rather than using self-centered evaluation of the situation. In the market actors behave based on self-centered calculation and do not take into consideration the expectation of others. In a hierarchy actors behave based on organizationally set rules and are controlled and coordinated by the chain of command. Mutual expectations, exercised in an open and fluid context, do not have an important place in bureaucratic organizations. In contrast, long-term relations are based on cooperation that can be realized only when interacting partners have common interests, understand mutual expectations, share resources, and enhance mutual gains. CCC interaction fits the Japanese institutional inheritance of interpersonalism.

This inheritance, however, is not the only factor influencing companies' choices. Coping with severe market competition and trying to survive, companies have to make interfirm relations flexible and efficient. The introduction of competition-generating measures serves this purpose; it makes CCC interfirm relations highly effective in responding to both the institutional inheritance and severe market competition.

Given the delicate problems of mixing cooperation and competition and facing the danger of making the wrong mix, companies use a set of norms and values to sustain the relationships. The four criteria of evaluation are: (1) whether CCC interfirm relations are developing cooperation and trust for generating benefits at present or for future; (2) whether partner companies are mutually sharing information and capabilities for survival; (3) whether partner companies are meeting the necessary criteria and demands for surviving in a competitive product market, including effective response to competition-generating measures; and (4) whether partner companies are risking themselves in setting up a future path of CCC interfirm relations.

This logic of continuity is the manifestation of Japanese interpersonalism. The institutionally inherited norms and values restrict the way partner companies interact, and force them to behave in ways to mutually satisfy expectations. Such restrictions could retard the dynamics in interaction, making partners incapable of coping with a turbulent corporate environment. CCC interfirm relations solve this problem by turning the restrictive elements of Japanese interpersonalism into dynamic ones. The importance of mutual expectations remains the same, but the expectations incorporate forward moving momentum for better performance. Companies set the goal of expectations in winning market competition cooperatively, raise standards to a very high level to ensure survival in the market, and transform standards from static factors to constantly moving targets that require frequent interaction in identifying goals. They incorporate competition-generating measures, though contradictory in nature to cooperation, as necessary means for satisfying mutual expectations and achieving commonly shared goals. And to ensure long-term relations for the future, they include future-oriented risk-taking activities as an important part of the logic of continuity. Although institutionally inherited norms and values restricted Japanese companies in the past, by reinterpreting the same principles to suit to the new corporate environment Japanese companies have transformed the old and restrictive institutional factors into new, flexible, and dynamic sources of interfirm relations. The outcome is the CCC interaction. This is why long-term relations even at present are a source of dynamic interaction in the Japanese semiconductor industry.

One best indication of the dynamics is that both types of competition- and cooperation-oriented CCC interaction work effectively in lowering prices. In cooperation-oriented CCC interaction, semiconductor companies emphasize trust, cooperation, and the human-relations-oriented logic of continuity with technologically inferior parts/material suppliers, and successfully achieve lower prices. They emphasize the performance-oriented logic of continuity with technologically superior partners, stringently apply competition-generating measures, and lead to the same result. And in competition-oriented CCC interaction of sales, technologically-sophisticated and highly attractive customers also achieve the same result with semiconductor companies. The introduction of a market-like environment in hierarchical relations also results in lowering prices. These findings suggest that CCC governance and hierarchical governance with a market-like environment are highly effective in lowering prices, much more than market and usual hierarchical relations.

The above findings further suggest that the market may not be the most effective institutional arrangement for generating the lowest price and achieving efficiency. CCC relations, whether cooperation- or competition-oriented, can achieve favorable prices to semiconductor companies and customers respectively in the procurement of parts and materials and in sales. In the end, a combination of cooperation-promoting and competition-generating measures, that results in a mixture of both allocative and non-allocative efficiency, may be more efficient than a simple market approach. Moreover, paths to achieve lower prices are diverse. Even highly human-relations-oriented behavior can achieve results equal to performance-oriented behavior. Even cooperation-oriented CCC relations can obtain results equal to competition-oriented CCC relations. What matters is a clear understanding of the conditions and the proper mix of cooperation-promoting and competition-generating measures.

With regard to the unfavorable price that technologically-inferior parts/material suppliers face, some may argue that the power asymmetry allows exploitation of smaller companies by large ones. The fact that the capability of semiconductor companies to price lower is not different between technologically-inferior and superior partners, may suggest that their argument is not applicable to this study. Moreover, considering the cost for semiconductor companies to develop the capability of technologically-inferior suppliers to the present level, the lower price may be considered as part of cooperative bargaining taking place under shared goals and asymmetric power conditions.

Thus, CCC interfirm relations delicately mix cooperation and competition, and generate comparative advantages that cannot be developed in neither hierarchical nor market governance. The CCC interfirm relations may not simply be an in-between form, but rather signify an institutional arrangement with distinctively different characteristics. A similar environment can also be developed by introducing a market-like environment in hierarchical relations. But CCC interaction works more effectively, when partners interact frequently, when influence is more mutual, when interaction is less organizationally restricted, and when the product market is more competitive. The capability of hierarchical governance to incorporate a market-like environment is limited.

Semiconductor manufacturers also use diverse types of partners, and generate the most dynamic effects by interacting with all different types. They make rational choices with whom they should interact on what business in what way. Even minor partners have their proper role to play within the overall structure of intra- and interfirm relations. I shall briefly identify how each type of partner is linked with others to maximize the dynamics of operations in semiconductor companies.

B.   Overall Structure of Intra- and Interfirm Relations

Table 6.1 identifies the overall structure of intra- and interfirm relations, based on average values of all indicators in flexible-synergy and market-like effects. (In the case of R&D, a mean value of Strict Standards and Tough

Table 6.1 Overall Structure of Intra- and Interfirm Relations

SALES	PROCUREMENT OF PARTS AND MATERIALS	R&D	PROCUREMENT OF MANUFACTURING EQUIPMENT
Core Elements to Dynamic Interaction			
		Hierarchical Governance (Internalized Interaction) Overall Flexible-synergy Effects = 4.10 (strong+) Strict Standards and Tough Relationship = 4.30 (strong+) Transactions = 91.8%	
Competition-oriented CCC Governance (Horizontal CCC Interaction) Overall Flexible-synergy Effects = 3.20 (medium+) Overall Market-like Effects = 3.51 (strong) Transactions = 50.6%	Cooperation-oriented CCC Governance (Vertical CCC Interaction) Overall Flexible-synergy Effects = 3.60 (strong) Overall Market-like Effects = 3.09 (medium+) Transactions = 33.1%		
Supportive Core Elements to Dynamic Interaction			
Hierarchical Governance (Internalized Interaction) Overall Flexible-synergy Effects = 3.35 (medium+) Overall Market-like Effects = 3.39 (medium+) Transactions = 19.0%	Cooperation-oriented CCC Governance (Horizontal CCC Interaction) Overall Flexible-synergy Effects = 3.38 (medium+) Overall Market-like Effects = 3.23 (medium+) Transactions = 47.9%		Cooperation-oriented CCC Governance (Horizontal CCC Interaction) Overall Flexible-synergy Effects = 3.38 (medium+) Strict Standards and Tough Relationship = 3.00 (medium+) Transactions = 5.4%
Competition-oriented CCC Governance (Vertical CCC Interaction) Overall Flexible-synergy Effects = 3.25 (medium+) Overall Market-like Effects = 3.26 (medium+) Transactions = 22.2%			

Relation is used as an indicator for market-like effects). The average value of a set of indicators for flexible-synergy or market-like effects is interpreted as showing the overall capability of both semiconductor companies and interacting partners to respond to cooperation-promoting or competition-generating measures respectively.

The structure of the Japanese semiconductor industry can be divided into core, supportive core, semi-peripheral, and peripheral elements, based on the average values of these two sets of indicators. Core elements have the capability of generating, the minimum, a medium+ level in one of two overall indicators and a higher level in the other. Supportive core elements have

Table 6.1 *Continued*

SALES	PROCUREMENT OF PARTS AND MATERIALS	R&D	PROCUREMENT OF MANUFACTURING EQUIPMENT
Semi-peripheral Elements to Dynamic Interaction			
	Hierarchical Governance (Internalized Interaction) Overall Flexible-synergy Effects = 3.33 (medium+) Overall Market-like Effects = 2.97 (medium) Transactions = 10.7%	Hierarchical Governance (Vertical CCC Interaction) Overall Flexible-synergy Effects = 3.70 (strong) Strict Standards and Tough Relationship = 2.60 (medium) Transactions = 2.2%	Cooperation-oriented CCC Governance (Vertical CCC Interaction) Overall Flexible-synergy Effects = 3.13 (medium+) Overall Market-like Effects = 2.93 (medium) Transactions = 25.5%
			Cooperation-oriented CCC Governance (Horizontal CCC Interaction) Overall Flexible-synergy Effects = 3.07 (medium+) Overall Market-like Effects = 2.80 (medium) Transactions = 48.4%
			Hierarchical Governance (Internalized Interaction) Overall Flexible-synergy Effects = 3.06 (medium+) Overall Market-like Effects = 2.74 (medium) Transactions = 13.2%
Peripheral Elements to Dynamic Interaction			
Market Governance (Spot Interaction) Overall Flexible-synergy Effects = 2.23 (weak+) Overall Market-like Effects = 2.97 (medium) Transactions = 8.2%	Market Governance (Spot Interaction) Overall Flexible-synergy Effects = 2.77 (medium) Overall Market-like Effects = 2.92 (medium) Transactions = 8.3%	Market Governance (Spot Interaction) Overall Flexible-synergy Effects = 2.05 (weak +) (weak+) Strict Standards and Tough Relationship = 2.00 (weak+) Transactions = 0.60%	Market Governance (Spot Interaction) Overall Flexible-synergy Effects = 2.30 (weak+) Overall Market-like Effects = 2.17 (weak+) Transactions = 12.9%

the capability of generating a medium+ level in both, while semi-peripheral elements can develop a medium level of overall market-like effects and a higher level of overall flexible-synergy effects. Peripheral elements have the capability to generate these effects below or equal to a medium level in both.

1.   Core Elements

Core elements are the most important for semiconductor companies. To meet their urgent needs, semiconductor companies choose a type of partner that generates the most suitable mix of flexible-synergy and market-like effects. They single out intrafirm relations for conducting R&D, concentrate their energy to sell their products to the most attractive horizontally-related customers, and comply with severe market conditions by cooperating with vertically-related parts/material suppliers.

Intrafirm relations, other than in sales, remain tightly controlled and co-ordinated, generate a strong sense of cooperation, and coordinate diverse actors effectively. Because of this capability, intrafirm relations tend to lack flexibility to implement competition-generating measures and develop a market-like environment. It becomes a distinct weakness in contrast to CCC interaction. However, because of this capability, hierarchical relations enable intrafirm actors to keep strict control over corporate secrets, strengthen their sense of trust and cooperation, exchange technological and R&D information, coordinate their activities smoothly, share human resources and equipment, implement good planning, and establish strict criteria for evaluation. This is why most of the R&D budget of semiconductor divisions is allocated within the company. Intrafirm interaction is the only available choice to satisfy these conditions.

The most dynamic example of competition-oriented CCC interaction is found in the interaction between horizontally-related customers and semiconductor companies in sales. The former are most important, but are the toughest, owing to their stable, bulk, and high-technology-oriented purchase of innovative products. The latter develop cooperative relations with the former in attempts to win contracts consecutively, and nurture pseudo-long-term relations. But semiconductor companies' cooperative behavior stimulates competition among themselves, increases the customers' demands for diverse services, and allows horizontally-related customers to greatly benefit from cooperative behavior, resulting in flexible-synergy effects. If they fail to meet these demands, another company will easily take over, since only a pseudo-long-term relation is operative. Severe market competition also allows horizontally-related customers to implement competition-generating measures, forcing semiconductor companies to sell products at quite unfavorable and stable prices, and to follow strict standards. The consequence is a mixture of flexible-synergy and market-like effects, despite the fact that semiconductor companies can at most develop pseudo-long-term relations in sales. It is the most dynamic model in competition-oriented CCC relations, because horizontally-related companies rely on the performance-oriented logic of continuity, make the best use of severe market competition, and use cooperation effectively as a means of achieving higher performance. Their interactive behavior matches best the mechanism whereby competition-oriented CCC governance harmonizes its diverse actors.

One of the most dynamic types of partners in cooperation-oriented CCC interaction is the vertically-related parts/material supplier. Even less technologically capable companies create business opportunities by pursuing different strategies from other types of partners, and become a core element of interactive dynamics. Since close cooperation between parts/material suppliers and semiconductor companies is necessary for coordinating delivery and quality control, vertically-related suppliers compete by maximizing the benefits of their cooperation with the latter. What makes them more attractive is the intensity of demands originating in horizontally-related customers. Responding favorably and quickly to the severe and diverse demands of semiconductor companies, vertically-related suppliers turn out to be preferred as cooperating partners to horizontally-related suppliers. Their dynamic and future-oriented interactive behavior, based on the norms of the human-relations-oriented logic of continuity, operates best under cooperation-oriented CCC governance and becomes a strong comparative advantage.

2.   Supportive Core Elements

Although supportive core elements are still very important part in the overall structure of intra- and interfirm relations, they are supportive only from the perspective of the dynamics in interaction. In sales, both vertically-related and intrafirm customers lighten sales efforts and increase sales stability. Horizontally-related parts/material suppliers, though they are dynamic partners in cooperation-oriented CCC interaction and sell close to the majority of parts and materials, are less preferred due to their weaker capability to generate flexible-synergy effects. In R&D horizontally-related companies are important partners for technological alliances, though their role is rather minor relative to internal R&D.

Due to their technological vulnerability, vertically-related customers tend to be dependent on semiconductor companies for technology and new product ideas. Their vulnerability enables the latter to develop strong human ties, cultivate a strong sense of cooperation, respond well to customer needs, continuously suggest new product ideas, develop future-oriented risk-taking activities, and secure sales. Their tendency to respond better to cooperation-promoting measures provides a favorable environment for semiconductor companies to stabilize sales, while it weakens their capability to implement competition-generating measures and take advantage of severe market competition. Their less matching behavior to competition-oriented CCC governance makes them a less dynamic type of partner in sales activities, hence, a supporting-core element of the overall structure in intra- and interfirm relations.

Unlike other intrafirm relations in the procurement of parts/materials and equipment, intrafirm sales relations involve dynamic interaction. They introduce a market-like environment within the organization, and display a complex mixture of flexible-synergy and market-like effects. An organizational structure functions to buffer the severity of competition from the market; it enables semiconductor and other divisions to engage in intensive cooperation and generate flexible-synergy effects. Even under the highly cooperative environment of hierarchical governance, these divisions, operated as independent profit centers, interact in a market-like environment, and successfully generate several market-like effects. Consequently, intrafirm customers demand the least favorable price and tough standards, while they involve a more equal power relationship, mutual influence, and dynamic cooperative interaction with the semiconductor division as members of the same company. They are easier partners to work with. Their demand for high-risk integrated circuits, though not large in proportion, also becomes significant in stabilizing sales. Hence, vertically-related and intrafirm customers are important in stabilizing sales for semiconductor companies.

Horizontally-related parts/material suppliers sell close to the majority of parts and materials to semiconductor companies, and their interaction with semiconductor companies is also a good example of cooperation-oriented CCC interaction. They respond especially well to competition-generating measures to adjust prices and maintain strict standards and tough relations. Despite this, semiconductor companies prefer to work with vertically-related suppliers, if feasible, since horizontally-related suppliers are less flexible and cooperative than vertically-related suppliers. Hence, from the perspective of the dynamics in interaction, horizontally-related suppliers remain a supportive core element.

Another supportive core element is alliances in R&D between semiconductor manufacturers and foreign or domestic horizontally-related companies, though alliances cover only a small part of the R&D budget. These companies take a division of labor over complementary technology with less frequent interaction than intrafirm R&D cooperation. They involve specifically defined contractual relationships, exchanges of mutually advantageous technologies, and generate a significant degree of flexible-synergy effects. Moreover, they evaluate each other in a strict manner over specifically defined goals. Since their relationship is specific, the alliances tend to be dissolved when the goals are achieved, making the relationship unstable.

3.   Semi-peripheral Elements

Semi-peripheral elements are a little less dynamic relative to core and supportive core elements. In the case of procuring manufacturing equipment, it is greatly due to less frequent interaction in the procurement process. In contrast, in the case of internal parts/material procurement and R&D cooperation with vertically-related companies, it is due to the less important role they play in the overall structure of intra- and interfirm relations.

Since the procurement of manufacturing equipment takes place less frequently, it generates much lower degrees of flexible-synergy and market-like effects. However, technological linkages involved in jointly developing firm-specific equipment secure long-term relations. Horizontally-related equipment suppliers cooperate in their business-related areas and respond effectively to semiconductor companies' competition-generating measures; but given their highly-sophisticated technological capability, they are less willing to develop future-oriented risk-taking projects in order to preserve their oligopolistic market position. While technological linkages assure long-term cooperation in equipment, the range of cooperation is limited and CCC interaction generates weaker effects than for parts and materials.

Vertically-related equipment suppliers are the best at generating both flexible-synergy and market-like effects, relative to other types of equipment suppliers. They receive diverse help from semiconductor companies to maintain their technological and managerial capability, develop a strong sense of human bonding, and even engage in an intermediate level of future-oriented risk-taking activities. But the help comes with the highly suppressive control of semiconductor companies, because they try to reduce dependence on powerful and oligopolistic horizontally-related equipment suppliers, especially in developing equipment that requires rather intensive interaction such as firm-specific testing equipment. Their interaction undeniably is cooperation-oriented CCC interaction, but their relations are tightly controlled, generating an environment similar to hierarchical governance.

In contrast to horizontally-related equipment suppliers, intrafirm equipment manufacturing units have a highly sophisticated technological capability and develop a strong sense of cooperation with semiconductor divisions. They play an important role in developing more risk-taking and high-cost equipment, for which organizational capability to effectively coordinate with semiconductor divisions becomes quite important. Intrafirm equipment manufacturing units, however, are disadvantaged in competing with horizontally-related equipment suppliers for highly advanced manufacturing equipment and with vertically-related suppliers for less advanced manufacturing equipment, due to incapability in delicately mixing flexible-synergy and market-like effects.

Although these three types of equipment suppliers are less dynamic in interaction relative to core and supportive core elements, they play an important role as suppliers of manufacturing equipment. Moreover, they establish a division of labor among themselves in a way that maximizes their market position, technological capability, and the effectiveness of governance under which they operate. Semiconductor companies' reliance on intrafirm equipment manufacturing units and their suppressive control over vertically-related equipment suppliers are because of needs for reducing their dependence and increasing their bargaining edge against horizontally-related oligopolistic equipment suppliers.

Interaction between semiconductor and vertically-related companies in R&D is less dynamic. Since the purpose of joint R&D is mostly to maintain the technological capability of vertically-related companies, it comes with semiconductor companies' strong control, generating an environment similar to hierarchical governance. The technological capability of vertically-related companies is so weak that joint R&D activities cannot help involving strong asymmetric control. This interaction, though less dynamic, is quite important to sustain the technological capability of vertically-related suppliers in parts/materials and equipment. It is this R&D support that enables semiconductor companies to flexibly respond to intensive and tough demands of horizontally-related customers and to bargain with oligopolistic and technologically-sophisticated horizontally-related equipment suppliers. Hence, even a semi-peripheral element is closely tied to core and other semi-peripheral elements.

The intrafirm procurement of parts and materials occupies a minor portion in business transactions. Semiconductor and intrafirm parts/material supply units nurture a stronger sense of trust and cooperation and keep corporate secrets, but hierarchical relations do not allow flexibility in introducing another supplier to create a market-like environment. Consequently, prices tend to be high. The role of intrafirm supply units is to cut procurement costs by making company-wide bulk purchases, importing parts and materials from affiliates in foreign countries, or supplying them internally. Whichever their involvement, the evidence does not seem to support their effectiveness, unless the intrafirm supply units provide very special parts or materials.

4.   Peripheral Elements

Peripheral elements, consisting of market governance in all functional areas, are not significantly contributing to the dynamics of the semiconductor industry. When a majority of transactions are conducted through CCC relations that require partners to cooperate and respond to competition-generating measures, companies in the spot market are severely disadvantaged. They are the least interested in cooperating with semiconductor companies and in generating flexible-synergy effects. The only time semiconductor companies interact with them is when they make a quick purchase of inexpensive supplies, a one-time purchase of expensive equipment or technology, or an extra profit by selling surplus in the spot market. They are not attractive in developing long-term relations, because they can hardly generate flexible-synergy and market-like effects. Their importance lies in the capability to make quick transactions.

Thus, semiconductor companies select different types of partners in each functional area. Looking at the overall structure of intra- and interfirm relations, these choices seem to be well coordinated in the ways they combine the benefits of diverse interactions. As core elements, semiconductor companies choose internal R&D simply for the most suitable R&D environment that intrafirm relations can provide, and try to maximize their sales to the most attractive horizontally-related customers. For coping with severe market competition and responding to severe and tough demands made by horizontally-related customers, vertically-related parts/material suppliers respond most favorably and effectively.

As supportive core elements, technologically-dependent vertically-related customers and highly cooperative intrafirm customers lighten sales efforts and increase sales stability. Horizontally-related parts/material suppliers, though they are important partners in CCC relations, tend to be less cooperative than vertically-related suppliers. Similarly, horizontally-related R&D partners tend to cooperate with semiconductor companies in specific areas, but their relations tend to be unstable.

Manufacturing equipment suppliers remain semi-peripheral elements in the dynamics of interaction due to the less frequent nature of procurement. But the firm-specific nature of jointly developed technology maintains long-term relations. Semiconductor companies, facing oligopolistic and technologically sophisticated horizontally-related equipment suppliers, try to create a more favorable environment for themselves by nurturing the capability of vertically-related equipment suppliers, and reducing their dependence on and strengthening their bargaining position against horizontally-related suppliers. Their joint R&D for keeping the technological capability of vertically-related companies up to date also helps to maintain their bargaining edge against horizontally-related equipment suppliers and their capability to cope with the severe demands of horizontally-related customers. Companies in the spot market are used only for quick and one-time purchases and sales. How semiconductor companies integrate diverse partners seems to be greatly influenced by the prevalence of competitive-cum-cooperative interfirm relations in the Japanese semiconductor industry.

Appendix

Appendix Table 2.1 Type of Semiconductor

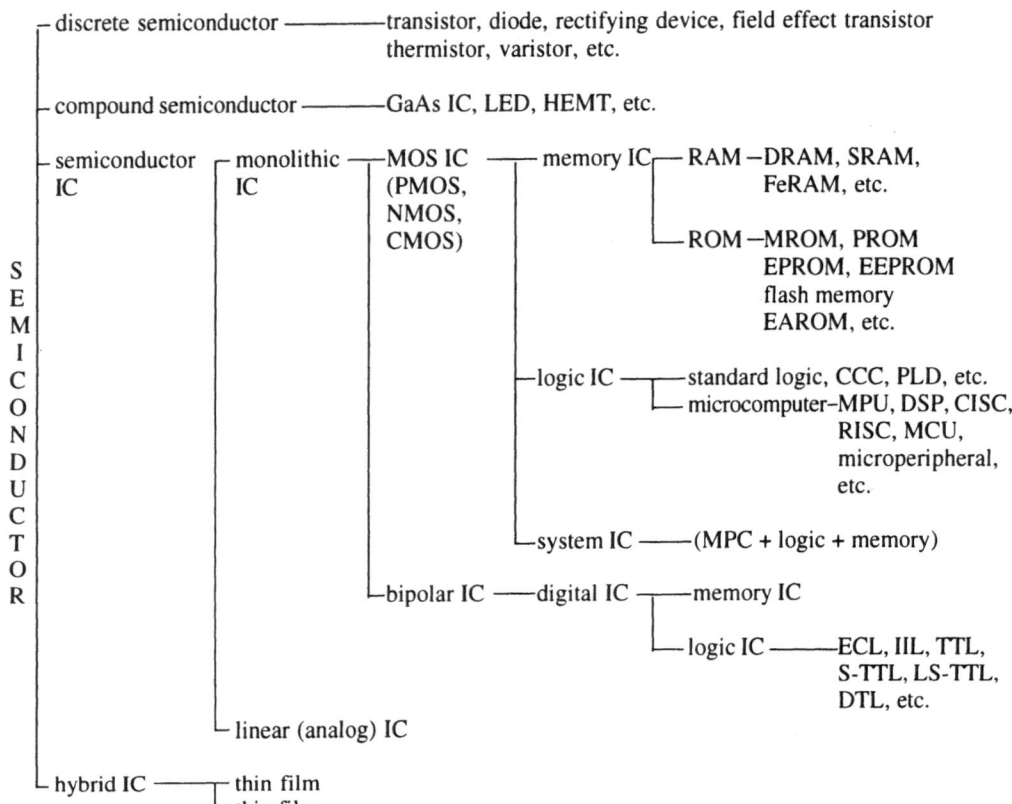

Note:
CCD = charge coupled device	CISC = complex instruction set computer	C-MOS = complementary-metal-oxide MOS
DRAM = dynamic random access memory		DSP = digital signal processor
DTL = diode-transistor logic	EAROM = electrically alterable ROM	ECL = emitter coupled logic
EEPROM = electrically erasable programmable ROM		EPROM = erasable programmable ROM
FeRAM = ferro-electric RAM	GaAs = gallium arsenide	HEMT = high electron mobility transistor
IC = integrated circuit	IIL = integrated injection logic	LED = light emitting device
LS-TTL = low power Schottky-TTL	MCU = micro controller unit	MOS = metal oxide semiconductor
MPU = microprocessor unit	MROM = mask ROM	N-MOS = N-metal-oxide MOS
PLD = programmable logic device	P-MOS = P-metal-oxide MOS	PROM = programmable ROM
RAM = random access memory	RISC = reduced instruction set computer	ROM = read only memory
SRAM = static random access memory		S-TTL = Schottky TTL
TTL = transistor transistor logic		

Source: United Nations Center on Transnational Corporations, Transnational corporations in the international semiconductor industry, (N.Y.: United Nations, 1986), 3; Nihon Denshi Kikai Kogyo Kai, '91 IC guidebook, (Tokyo: Nihon Denshi Kikai Kogyo Kai, 1991), 103-108; Chunichisha, Denshi buhin nenkan 1996/1997 (Annual of electronic device and components), (Tokyo: Chunichisha, 1997); and Press Journal, Nihon handotai nenkan (Japan semiconductor yearbook), (Tokyo: Press Journal, 1996), 362-363.

Appendix Table 2.2 Technological Development and Rough Estimates of Time Lags in Selected Semiconductor Products

Year	Transistors	Integrated Circuits (Excluding Microprocessors)	Microprocessors, Frontier Products, and Others
1947	U: Bardeen & Brattain: Ge point contact transistor: Success in experiment (Bell) @		
1948	U: Ge point contact transistor (WE) @ U: Shockley: Ge PN junction transistor (WE) @		
1949	U: PN Junction Transistor; Theoretical completion (Bell) @ U: PN Junction by diffusion method (WE) @		
1950	U: Hall; PN junction by alloying method (GE) @		
1951	U: Shockley, Sparks, & Teal: PN junction transistor; Completion of PN junction technique (Bell) @ J: Point contact transistor (TRI) #		
1952	U: Hall; Transistor by rate grown method (GE) @ U: Dacey & Ross; PN junction field effect transistor (FET); Basic technology leading to MOSFET in 1960 (Bell) @ U: Production of PN junction transistor for military purposes @ J: PN junction transistor (TRI) # U: PN junction transistor with alloying method (GE) >		
1953	U: Transistor used for a hearing aid: First use for a consumer product @ U: Silicon transistor (TI) > J: PN junction transistor with alloying method (Hitachi, SONY) >		
1954	U: Transistor radio (Regency); 6 months earlier than SONY # J: Production of transistor for consumer goods @ J: Transistor radio (SONY) @		
1955	J: Production of PN junction transistor by rate grown method (SONY) #		
1956	U: Lee; Mesa transistor (Bell) @		
1957	J: Esaki; Esaki diode (SONY); World first @ B: Beale; Alloy diffused transistor @ J: Production of mesa transistor (major companies) #	U: Molectronics (WH) @ U: Micromodule (RCA) #	
1958	J: Production of alloy diffused transistor (Matsushita) # UJ: Production of all transistorized computer @		

Time lag estimates (left margin): 4, 1, 3, 0.5, 1, 1, 0

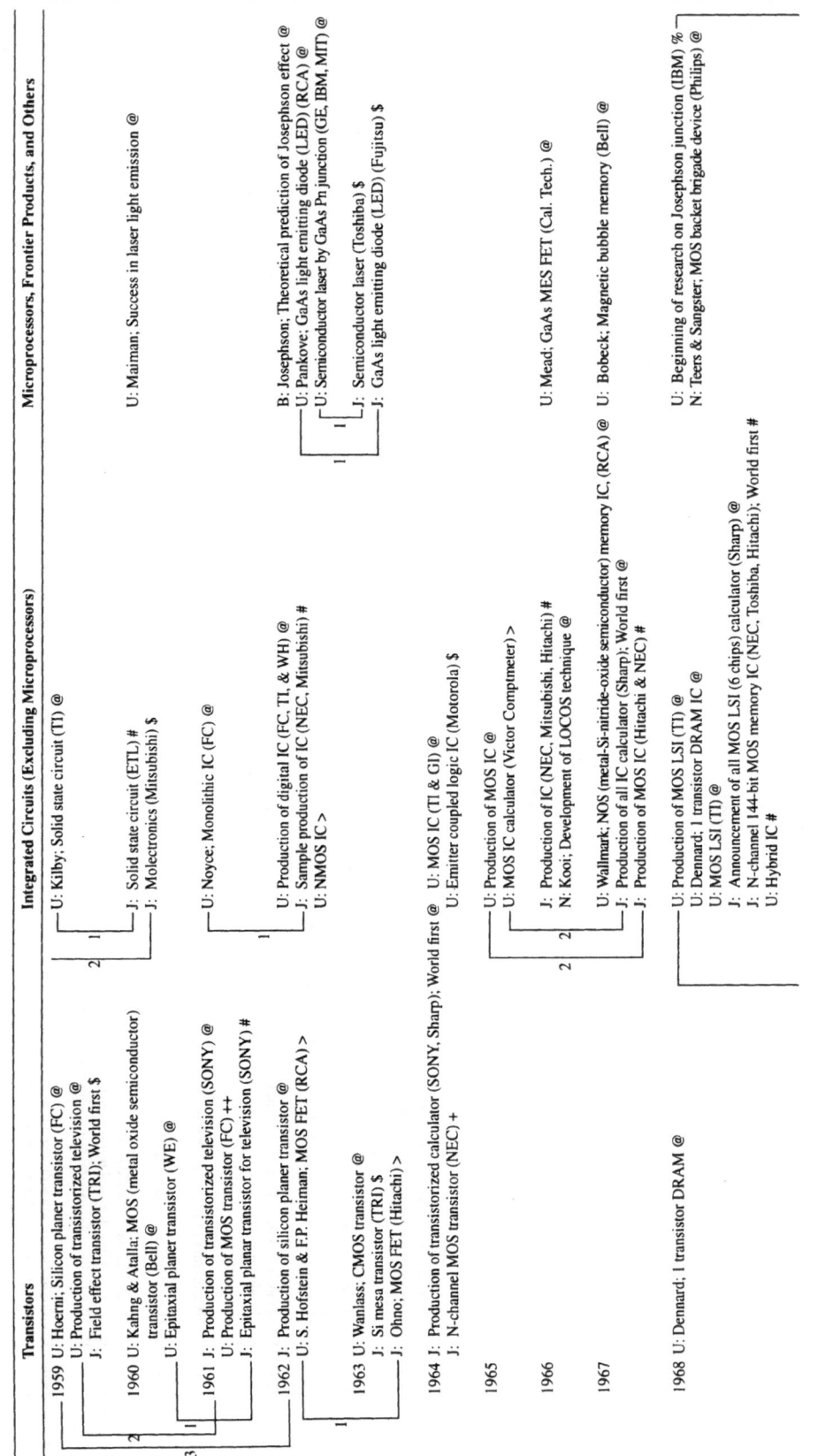

Year	Transistors	Integrated Circuits (Excluding Microprocessors)	Microprocessors, Frontier Products, and Others
1959	U: Hoerni; Silicon planer transistor (FC) @ U: Production of transistorized television @ J: Field effect transistor (TRI); World first $	U: Kilby; Solid state circuit (TI) @	
1960	U: Kahng & Atalla; MOS (metal oxide semiconductor) transistor (Bell) @ U: Epitaxial planer transistor (WE) @	J: Solid state circuit (ETL) # J: Molectronics (Mitsubishi) $	U: Maiman; Success in laser light emission @
1961	J: Production of transistorized television (SONY) @ U: Production of MOS transistor (FC) ++ J: Epitaxial planar transistor for television (SONY) #	U: Noyce; Monolithic IC (FC) @	
1962	J: Production of silicon planer transistor @ U: S. Hofstein & F.P. Heiman; MOS FET (RCA) >	U: Production of digital IC (FC, TI, & WH) @ J: Sample production of IC (NEC, Mitsubishi) # U: NMOS IC >	B: Josephson; Theoretical prediction of Josephson effect @ U: Pankove; GaAs light emitting diode (LED) (RCA) @ U: Semiconductor laser by GaAs Pn junction (GE, IBM, MIT) @ J: Semiconductor laser (Toshiba) $ J: GaAs light emitting diode (LED) (Fujitsu) $
1963	U: Wanlass; CMOS transistor @ J: Si mesa transistor (TRI) $ J: Ohno; MOS FET (Hitachi) >	U: MOS IC (TI & GI) @ U: Emitter coupled logic IC (Motorola) $	
1964	J: Production of transistorized calculator (SONY, Sharp); World first @ J: N-channel MOS transistor (NEC) +		
1965		U: Production of MOS IC @ U: MOS IC calculator (Victor Compumeter) >	
1966		J: Production of IC (NEC, Mitsubishi, Hitachi) # N: Kooi; Development of LOCOS technique @	U: Mead; GaAs MES FET (Cal. Tech.) @ U: Bobeck; Magnetic bubble memory (Bell) @
1967		U: Wallmark; NOS (metal-Si-nitride-oxide semiconductor) memory IC, (RCA) @ J: Production of all IC calculator (Sharp); World first @ J: Production of MOS IC (Hitachi & NEC) #	
1968	U: Dennard; 1 transistor DRAM @	U: Production of MOS LSI (TI) @ U: Dennard; 1 transistor DRAM IC @ U: MOS LSI (TI) @ J: Announcement of all MOS LSI (6 chips) calculator (Sharp) @ J: N-channel 144-bit MOS memory IC (NEC, Toshiba, Hitachi); World first # U: Hybrid IC #	U: Beginning of research on Josephson junction (IBM) % N: Teers & Sangster; MOS backet brigade device (Philips) @

Year	Transistors	Integrated Circuits (Excluding Microprocessors)	Microprocessors, Frontier Products, and Others
1969	J: DSA MOS transistor (ETL) >	U: Faggin, Klein, & Vadesz; Si-gate MOS IC > U: CMOS LSI (RCA) ++	
1970		U: Berger & Wiedmann; Integrated injection logic IC, (IBM) @ J: Production of MOS LSI (Hitachi, NEC, & Toshiba) # J: CMOS LSI (Toshiba) # J: 2048-bit MOS memory LSI (Hitachi); World first $ N: LOCOS IC (Philips) > U: 1K-bit SRAM IC (Intel) $ U: Boyle; Charge coupled device (Bell) @	
1971	J: Static induction transistor (Nishizawa) >		U: 4-bit microprocessor (Intel) @
1972		J: 1K-bit N-channel MOS memory LSI (one chip memory)(NEC) # U: 1K-bit N-channel MOS memory LSI (IBM) > U: EPROM IC (Intel) >	U: 8-bit microprocessor (Intel) ++
1973		J: 4K-bit N-channel MOS memory LSI (one chip memory) (Hitachi) #	J: 4-bit microprocessor (NEC) &
1974			U: GaAs digital IC (HP) @ U: 16-bit microprocessor (NSC) ++ J: 16-bit microprocessor (NEC) $
1975			J: 8-bit single chip microcomputer (Hitachi) > U: 8-bit single chip microcomputer (Intel) ++ J: Beginning of research on Josephson junction $
1976		J: 16K-bit MOS memory IC (Hitachi) $ U: 16K-bit MOS memory IC (Intel) ++	
1977		J: 64K-bit MOS memory IC (TRI): World first >	U: Archer; 1M-bit magnetic bubble memory (Rockwell) @
1978		J: 128K-bit MOS memory VLSI (TRI); World first $ J: 16K-bit CMOS RAM IC $	U: Concept of high electron mobility transistor (Bell) $
1979			U: Signal processor (Intel) @ J: 16-bit single chip microcomputer (TI, IBM) $ J: 16-bit single chip microcomputer (Toshiba) $ U: McAlear; Concept of biochip "Molton" @

	Transistors	Integrated Circuits (Excluding Microprocessors)	Microprocessors, Frontier Products, and Others
1980		J: 256K-bit DRAM IC (NEC, Toshiba, TRI); World first @ U: EEPROM IC (Intel) @ -2	J: High electron mobility transistor (HEMT) (Fujitsu); World first @
1981			U: 32-bit microprocessor (Intel, HP) @
1982		U: 256K-bit DRAM IC (IBM. Bell) >	J: GaAs 1K-bit SRAM (TRI); World first % U: Production of 32-bit microprocessor (Intel) & -2
1983			U: IBM stopped Josephson junction research (IBM) % U: GaAs MESFET 1K-bit gate array (Textron) % J: Josephson 1K-bit memory (TRI); World first !
1984		J: 1M-bit DRAM IC (TRI, NEC, Hitachi, Fujitsu, & Toshiba) @ U: 1M-bit DRAM IC (IBM) $ J: 256K-bit SRAM IC (TRI, NEC, Hitachi,Fujitsu, Toshiba); World first @ J: PLL LSI (NPC) ## J: Three dimension memory LSI (NEC, Mitsubishi); World first ! J: 1M-bit EPROM IC (NEC); World first ! J: 256K-bit SRAM IC (Toshiba); World First ##	J: HEMT 1K SRAM (Fujitsu); World first % J: GaAs 4K-bit SRAM (Fujitsu, TRI, Toshiba) @ U: GaAs 1K-bit SRAM (Rockwell) % J: Josephson chip for multiplication (TRI); World first ! J: GaAs MES FET 1K-bit gate array (Toshiba) % J: 1-chip bio sensor (NEC); World First % U: 32-bit microprocessor (Hitachi) $ U: Heterojunction bipolar transistor (Rockwell & TRI) !

All listings after 1985 are "world first" developments.

	Transistors	Integrated Circuits (Excluding Microprocessors)	Microprocessors, Frontier Products, and Others
1985		J: CMOS 512K-bit EPROM IC (Fujitsu) ##	J: GaAs logic LSI (NEC) ^ J: 1K-bit lead-based Josephson memory (NTT) =
1986		J: 4M-bit DRAM IC (NEC, Toshiba) ^ U: 4M-bit DRAM IC (TI) * J: 8K-bit SRAM three dimension LSI (Matsushita) ##	U: RISC Processor (MIPS) %
1987		J: 16M-bit DRAM IC (NTT) ^ J: 2M-bit EPROM IC (NEC) ## J: 8M-bit mask ROM IC (Matsushita) ## J: 16M-bit mask ROM IC (Matsushita) ##	J: HEMT 4K-bit SRAM (Fujitsu) ^ J: Single phase yttrium barium copper oxide Josephson device (NEC) @@
1988		J: 4M-bit CMOS EPROM IC (NEC) ## J: Bi-CMOS 128K-bit PROM IC (Fujitsu).##	J: Josephson junction microprocessor (Fujitsu) ## J: GaAs hetero bipolar transistor with complete monolithic IC (NEC) ## J: 1K-bit optical memory with vertical to surface transmission electron phontonic device (NEC) ##

Transistors	Integrated Circuits (Excluding Microprocessors)	Microprocessors, Frontier Products, and Others
1989	J: 16M-bit EPROM IC (NEC) ## J: 4M-bit EEPROM IC (Toshiba) &	J: Josephson clock conversion circuit (Fujitsu) ## J: Josephson computer (NTT) @@ U: 32-bit GaAs RISC microprocessor (TI) ## J: 4K-bit Josephson memory (Fujitsu) ## U: 64-bit microprocessor (Intel) ##
1990	J: 4M-bit SRAM IC (Hitachi, Toshiba, NEC, SONY) $$ J: 64M-bit DRAM IC (Hitachi) $$ J: 16M-bit CMOS EPROM IC (NEC) %% J: Low temperature BiCMOS IC (NEC) ** J: 4M-bit CMOS transistor-transistor logic SRAM IC (Fujitsu) ##	J: Josephson 8-bit DSP (Fujitsu) ## J: HEMT 64K-bit SRAM (Fujitsu) ##
1991		U: GaAs computer (Convex Computer) && U: Semicustom transputer (Inmos) ^^

Note: B: Great Britain   J: Japan   K: Korea   N: Netherlands   U: U.S.   Bell: Bell Telephone Laboratory   Cal. Tech.: California Institute of Technology
ETL: Electro-Technical Laboratory (Denkishikenjo – Currently Denshi Gijutsu Sogo Kenkyusho in MITI)   FC: Fairchild Camera & Instrument Corp.   GE: General Electric Co.
GI: General Instrument   HP: Hewlett-Packard Co.   MIT: Massachusetts Institute of Technology   NEC: Nippon Electric Co., Ltd.   NPC: Nihon Precision Circuit (Seiko Group)
NSC: National Semiconductor Corp.   TI: Texas Instruments Inc.
TRI: Telecommunication Research Institute (Denkitsushin Kenkyusho in NTT [Nihon Denshin Denwa Kosha, indicated as NTT after 1985])   WE: Western Electric Co. Inc.
WH: Westinghouse Electric Corp.

Source: @ Shunkichi Shirosaka, *Kagaku gijutsushi* (History of scientific technology). (Tokyo: Nikkan Kogyo Shinbunsha, 1984).
$ Denpa Shinbunsha, *Denshi kogyo nenkan* (Yearbook of electronics industry). (Tokyo: Denpa Shinbunsha, 1964-1991).
# Yasuzo Nakagawa, *Nihon no handotai kaihatsu* (Development of semiconductors in Japan), (Tokyo: Diamond, 1985).
^ Nihon Denshi Kogyo Shinko Kyokai (Japan Electronic Industry Development Association), *Denshi kogyo shinko 30 nen no ayumi* (Thirty-year history of electronics industry development), (Tokyo: Nihon Denshi Kogyo Shinko Kyokai, 1988).
% ____, "*Shindenshi zairyo ni kansuru chosa kenkyu hokoku sho XI: Gijutsu joho chosa hokoku* (Research report on new materials in electronics industry: Technical report)," (Tokyo: Nihon Denshi Kogyo Shinko Kyokai, 1985).
& Press Journal, *Nihon handotai nenkan* (Japan semiconductor yearbook), (Tokyo: Press Journal, 1985-1992).
! Piko Nano, *Haiteku suizensen no yomikata* (How to read the forefront of high-technology), (Tokyo: Ko Shobo, 1985).
= Marie Christine Anchordoguy, The state and the market: Industrial policy toward Japan's computer industry, Ph.D. Thesis, University of California-Berkeley, 1986.
@@ Martin Fransman, The market and beyond: Cooperation and competition in information technology development in the Japanese system, (Cambridge: Cambridge University Press, 1990).
Sangyo Times, *Handotai sangyo keikaku soran* (Comprehensive list of semiconductor industry planning). (Tokyo: Sangyo Times, 1985-1991).
* Nihon Keizai Shinbun, "IC sangyo futatabi kosei e (IC industry booming again)," Sept. 17, 1986, 2nd Section.
$$ ____, "*64M DRAM shisaku seiko* (Success in developing 64M DRAM in experiment)," June 8, 1990.
%% Nikkan Kogyo Shinbun, "*Nihondenki, 16M CMOS EPROM o kaihatsu* (NEC develops 16M CMOS EPROM)," February 14, 1990, 8.
** ____, "*Nichiden shinkairo kaihatsu* (NEC develops new circuit)," Dec. 21, 1990.
++ United Nations Centre on Transnational Corporations, Transnational corporations in the international semiconductor industry, (New York: United Nations, 1986), 455-456.
> *Handotai gijutsu no kaihatsu nenpyo* (Chronology of semiconductor development), unpublished document, Hitachi, Tokyo.
+ Unpublished report provided to the author by Company B.
&& Electronic-World-News, "Convex goes with GaAs for C3800," May 20, 1991, 14.
^^ Electronics-Times, "Inmos delivers first customized transputer," November 14, 1991.

Appendix Table 2.3 Selected Semiconductor-industry-related Events In Japan

1948 J: Electro-Technical Laboratory (*Denkishikenjo*) of the Ministry of Communica-
         tion (*Teishinsho*) split into the Telecommunication Research Institute (TRI:
         Denkitsushin Kenkyusho) of the Ministry of Telecommunication
         (*Denkitsushinsho*) and the Electro-Technical Laboratory (ETL: *Denkishikenjo*)
         of the Ministry of Commerce and Industry (*Shokosho*).
     J: Private study group on the transistor was formed in ETL.

1949 J: Government research grant acknowledged the importance of the study group,
         which thereafter named itself the Transistor Study Group.
     J: TRI began research on the transistor.

1950 J: Watanabe and Nishizawa of the Tohoku University obtained a patent on the
         technology formulating I layer with ion implantation and also on the intrinsic
         semiconductor theory.

1951
1952 UJ: Technical agreements: Hitachi, Toshiba and Kobekogyo (later acquired by
         Fujitsu) with RCA on the transistor.
     EJ: Technical agreement: Fuji Electric with Siemens on the transistor.
     J: Patent on the multiplex diffusion technology (TRI).
     U: Patent on the jet etching process technology (Philco).
     J: Nippon Telegraph and Telephone Public Corporation (*Nihon Denshin Denwa
         Kosha*) was established, and TRI became its research institute.

1953 UJ: Technical agreement: SONY with Western Electric (WE) on the transistor.
     J: Watanabe and Nishizawa of the Tohoku University obtained a patent on the pin
         rectifier technology.

1954 UJ: Technical agreements: Toshiba, Hitachi, and Kobekogyo with WE on the
         transistor.
     J: Patent on the surface processing technology (SONY).

1955 U: Patents on the oxide masking and diffusion technologies (Bell Telephone
         Laboratory: Bell).

1956 J: Mass-production of transistor started by the first group (SONY, NEC, Toshiba
         and Hitachi).

1957 J: Electronics Industry Promotion Law (*Denshi Kogyo Shinko Rinji Sochi Ho*:
         Denshinho) and the Machinery Industry Promotion Law (*Kikai Kogyo Shinko
         Rinji Sochi Ho*: *Kishinho*) were promulgated.
     J: Patent on the variable capacitance diode (TRI).
     UJ: Technical agreement: SONY and RCA on the transistor.
     EJ: Technical agreement: Matsushita and Philips on the transistor.

1958 UJ: Technical agreements: NEC with RCA, SONY with General Electric (GE),
         Fuji Electric with RCA and WE, and NEC with GE on the transistor.

1959 J: Second group of companies entered the transistor market (Mitsubishi Electric, Japan Radio, Oki Electric and Sanyo).

UJ: Technical agreements: Sanyo, Oki Electric and Japan Radio with WE and RCA on the transistor and diode; and Mitsubishi Electric with RCA on the transistor.

J: Japan became the world largest transistor producer, and the U.S. started restricting the import of Japanese transistor.

J: Quality of Japanese rate grown transistor surpassed that of U.S.

J: First transistorized television (SONY).

J: First 100% domestic-component transistor television was developed (Toshiba).

J: First 100% domestic-component color television was developed (Sharp).

J: First computer with transistors was developed (NEC).

U: U.S. Government began an inquiry on the threat of Japanese transistor import to the national security.

J: Hitachi established the Transistor Laboratory.

1960 J: Research on the IC began (NEC and Mitsubishi Electric).

J: Quota was set for transistor export.

UJ: Technical agreements: Mitsubishi Electric with Westinghouse (WH), leading to the announcement of molectronics; and Shindengen Electric with WE, Sanken Electric with RCA, and Origin Electric with WE on the transistor and diode.

1961 J: Boom in establishing research institutes.

J: Tohoku University established a research institute on the IC.

J: Patent on the power switching transistor (Hitachi).

UJ: Technical agreements: Hachio Electric (currently Fujitsu General) with WE, and Oki Electric with General Instrument (GI) on the transistor and diode.

1962 J: Thirty researchers moved from TRI to private companies.

J: NEC became the sole representative to administer Fairchild's (FC) planer patent in Japan.

J: Patent on the high pressure resistant diode (NEC).

J: NEC and Mitsubishi Electric succeeded in making IC samples, but found no market.

J: Japanese Government provided a large research grant to develop a high capability calculator "FONTAC" (1962-64) (FONTAC *Kaihatsu Hojokin*).

U: Flat pack packaging technology was invented (Texas Instruments: TI).

J: Increasing number of small- and medium-sized companies became independent from subcontracting because of their improved technical capability.

UJ: Technical agreement between Meidensha and WE on the transistor and diode.

1963 J: Patents obtained on the low temperature passivation (LTP), two other processing technologies (Hitachi) and the pinch resistor for IC (NEC).

UJ: Technical agreements: Toshiba and Shindengen Electric with GE on the unijunction transistor; SONY, Hitachi, Mitsubishi Electric, Oki Electric and Sharp with WE, New Japan Radio with Raytheon, and SONY with GE on semiconductor manufacturing equipment; Shindengen Electric with Clevite, NEC with WE, and Sansha Electric with RCA on the transistor; Toshiba and Sanken Electric with Clevite on the high power switching transistor; NEC with FC on the planer transistor and micro-logic unit; and Japan Storage Battery with WE and RCA on the semiconductor.

1964 U: TI attempted to obtain a foreign investment permission from the Japanese
         Government.
     J: Fujitsu acquired Kobekogyo.
     J: SONY licensed the Esaki diode production technology to IBM, RCA and TI.
     U: Manasevit developed the silicon-on-sapphire technology.
     U: Solid logic technology for complete production automation was developed
         (IBM).
     U: Dual-in-line packaging technology was invented (FC).
     U: Flip-chip-method packaging technology was invented (IBM).
     UJ: Technical agreements: Kobekogyo, Hitachi and Mitsubishi Electric with
         Clevite on the high power switching transistor.

1965 U: TI opened its patents to the public in Japan.
     J: NEC announced its strong commitment to IC development.
     J: Patents on several technologies for metal oxide passivation (MOP) (Mitsubishi
         Electric), multiple masking (Hitachi), MOS transistor protection (Toshiba),
         $SiSiO_2$ metallic oxide filming (Mitsubishi Electric), and LOCOS processing
         (NEC).
     JJ: Joint development: Nippon Telegraph and Telephone Public Corporation with
         four private companies on electronic telephone switches.
     EJ: Technical agreement between Fuji Electric and Siemens on the transistor.
     UJ: Technical agreements: Origin Electric, Shindengen Electric and Fujitsu with
         WE, Stanley Electric with WE on semiconductor manufacturing equipment;
         New Japan Radio with RCA on the transistor; Origin Electric with RCA on the
         high power switching transistor; and Hitachi with RCA on the IC.

1966 J: Patent on the $SiSiO_2$ alminophosphide glass surface processing technology
         (Hitachi).
     J: Electron beam printing technology (ETL) and the IC for computer were
         developed (Hitachi, NEC, and Fujitsu).
     J: Japanese Government established a project (1966-1971) to develop a high
         capability computer that would be competitive against the IBM 360 computer
         (*Chokoseino Denshi Keisanki Purojekuto*).

1967 J: Patents on an ion implantation method (Hitachi) and the collector for
         polycrystalized Si (SONY).
     U: Patent on an ion implantation method (Bell Telephone Laboratory).
     J: Japan Electron Optics Laboratory (Nihon Denshi) developed the world's first
         electron beam lithography machine with ETL.
     J: Large number of bankruptcies among electronic parts producers.
     UJ: Technical agreements: Hitachi, Toshiba, Mitsubishi Electric and Fujitsu with
         FC on the IC.
     JJ: Joint development: Nippon Telegraph and Telephone Public Corporation with
         NEC on a telephone switching system with ICs.

1968 UJ: TI established Texas Instruments Japan, and formed a joint venture with
         SONY.
     J: Inexpensive U.S. ICs flooded the Japanese market.
     J: Patents obtained on the inverted ion implantation (ETL), ion implantation
         metalization (Hitachi), and channel stepper technologies (NEC).

J: Government lifted control over technical agreements for IC.

UJ: Technical agreements: Toshiba, NEC, Hitachi, Mitsubishi Electric, SONY and
       TI Japan with TI on the semiconductor; SONY, Toshiba and Hitachi with
       Canada GE on the alloy transistor; Sharp with WE on semiconductor manufac-
       turing equipment; and Toshiba with RCA on the IC.

1969 J: Sharp imported MOS LSIs for calculator from the U.S.

    J: Patent on the DSA MOS transistor (ETL).

    U: Patent on the collector diffusion isolation technology (Bell Telephone Labora-
           tory).

    UJ: Technical agreements: New Japan Radio with TI on the semiconductor and IC;
           Nippon Columbia and Mitsumi Electric with FC on the planer semiconductor
           and IC; Kokusai Electric with TI, and Toyo Electric Seizo with WE on semi-
           conductor manufacturing equipment; Tokyo Sanyo with GI on the MOS IC;
           Asahi Glass with Corning Glass Works on the package-type IC; Nippon
           Columbia with Integrated Circuit Systems on the MOS LSI gate array; and
           Nippon Columbia with TI, and Mitsubishi Electric with TRW on the IC.

1970 U: Importation of low-priced U.S. ICs from developing countries increased.

    U: Quality problems of U.S. ICs gave a comparative advantage to Japanese
           producers.

    J: Toshiba developed the perfect crystal (PCT) and chemical vapor deposition
           (CVD) technologies and manufacturing equipment, which relieved the com-
           pany from dependence on the planer patent.

    J: Low-powered high-capacity LSI was developed (ETL).

    U: U.S. pressured to open the computer market during textile trade disputes.

    U: Patent on the silicon-on-sapphire technology (SOS) (RCA).

    J: Hitachi opened all of its patents to the public.

    J: Electro-Technical Laboratory (*Denkishikenjo*) was reorganized into the Electro-
           Technical Laboratory (*Denshi Gijutsu Sogo Kenkyusho*) in the Ministry of
           International Trade and Industry.

    UJ: Technical agreements: Toshiba with Dow Corning on silicon manufacturing
           equipment; Nippon Columbia, SONY, Shindengen Electric, Origin, Toshiba,
           Mitsumi Electric, and Hitachi with WE on semiconductor manufacturing
           equipment; Mitsubishi with WE on the semiconductor and manufacturing
           equipment; NEC with FC on the planer semiconductor and IC; Mitsubishi
           Monsanto with Monsanto, Sansha Electric with GE, Matsushita Electronics
           with FC, and Fujitsu and Toko with WE on the semiconductor; Yamaha with
           Philco on the MOS IC for electronic music instrument; and Hitachi and Sharp
           with TI on the IC.

1971 U: Major U.S. electronics companies withdrew from memory IC production.

    U: Calculators with imported U.S. ICs and LSIs malfunctioned frequently,
           damaging their manufacturers' reputations.

    J: CMOS LSI for hand calculator was developed (Toshiba).

    UJ: TI resolved the joint venture with SONY.

    J: Industrial Promotion Law for Specific Electronics and Machinery Industries
           (*Tokutei Denshi Kogyo Oyobi Tokutei Kikai Kogyo Shinko Rinji Sochi Ho:
           Kidenho*) took over the previous *Kishinho* and *Denshinho*.

J: Foreign ownership of up to 50% was allowed for companies producing ICs with less than 50 elements.

JJ: Hitachi and Fujitsu cooperated in computer development, and established the *Cho Koseino Konpyuta Kaihatsu Gijutsu Kyodo Kumiai* (Very High Performance Computer Technology Development Cooperative).

JJ: Toshiba and NEC cooperated in computer development, and established the *Shin Konpyuta Sirizu Gijutsu Kenkyu Kumiai* (New Computer Series Technology Research Cooperative)."

JJ: Mitsubishi Electric and Oki Electric cooperated in computer development, and established the *Cho Koseino Denshi Keisanki Gijutsu Kenkyu Kumiai* (Very High Performance Computer Technology Research Cooperative).

J: Canon entered the semiconductor market.

UJ: Technical agreements: Matsushita Electronics, Toshiba and TI Japan with TI on the IC; Shindengen Electric with International Standard, New Japan Radio with Raytheon, and Toshiba and Oki Electric with WE on the semiconductor; Sharp with North American Rockwell on the MOS LSI; and Sanken Electric with GE on the diode.

EJ: Technical agreements: Hitachi with Philips on the germanium semiconductor; Toshiba with Philips on the semiconductor; and Sanyo with Telefunken on the transistor.

1972 J: Increased production automation and enhanced quality control: a completely automated production system (NEC), the automated LSI mask defect detection and ion implantation technologies, a mass production technology with tape careers (Sharp), and high quality materials for ICs (Shinetsu Handotai).

UJ: Fairchild established Nippon Fairchild in Japan.

J: Government established a research grant program for developing computers and related technologies (1972-1976) (*Denshi Keisankito Kaihatsu Sokushinhi Hojokin*).

J: Voluntary restriction on the export of electronic calculator.

UJ: Joint venture between TDK and FC for semiconductor production.

UJ: Technical agreements: Sanyo with WE, and TDK Fairchild and NEC with FC on the semiconductor; Daini Seikosha with Micropower System on germanium manufacturing equipment; Toko with Advance Memory Systems, and Mitsubishi Electric with GE and WH on the IC; and NEC with GE on the diode.

EJ: Technical agreements: NEC, Fujitsu, Sanyo, Mitsubishi Electric, and SONY with Philips on germanium manufacturing equipment.

1973 J: LSI needed for microprocessor was developed (NEC).

J: Oil Shock.

U: Non-contact direct optical injection system was developed (Perkin Elmer).

UJ: Motorola established Nippon Motorola in Japan, and the world top three IC producers were in production in Japan.

UJ: Joint venture between Alps Electric and Motorola for semiconductor production.

UJ: Technical agreements: Toshiba with GE, and NEC with FC on the IC and semiconductor; Mitsubishi Electric and SONY with International Standard Electric, and Toshiba with GE on the transistor; Kishimoto Sangyo with Pacific

Resins and Chemicals on the light emitting diode; Alps Motorola Semiconductor with Motorola on the IC and LSI; Hitachi and Sanyo with RCA, and Hitachi Chemical with Kollmorgen on the IC; Hitachi with FC, NEC with Motorola, and Sharp with WE on semiconductor manufacturing equipment; and Mitsubishi Electronics and Toshiba with FC, and Matsushita Electronics with RCA on the semiconductor,

EJ: Technical agreement between Osaka Titanium and Siemens on the semiconductor.

1974*J: 100% foreign ownership was allowed for a company producing ICs except for computer use.

J: Government lifted control over IC import.

J: TRI started the first VLSI development project (1975-1977).

J: Nixon Shock.

U: Electron beam exposure system was developed.

J: World's first reactive ion etching equipment (Nichiden Anelva).

1975*J: Advancement in the chemical dry etching (CDE) technique (Toshiba Tokuda Seisakusho).

JJ: Joint development: Fujitsu, Sharp, and Kyodo Electronics on the linear IC for industrial use.

J: Nippon Denso entered the semiconductor market.

UJ: U.S. investment, ITT Semiconductors, was established.

1976 J: Government established a research cooperative for developing VLSI technologies (1976-1979) (*Cho* LSI *Kaihatsu Sokushinhi Hojokin*).

UJ: Technical agreements: Nippon Denso with RCA on the IC; Hitachi with American Microsystems and MOS Technology on the semiconductor and IC and manufacturing equipment; New Japan Radio with Raytheon on the semiconductor; and Toko with WE, Hitachi with Intel and WE, and NEC with Intel on semiconductor manufacturing equipment.

UJ: Intel established Intel Japan for designing ICs.

1977 J: IC trade disputes with the U.S. started.

J: Pioneer started producing its own semiconductors.

UJ: Technical agreements: Hitachi with RCA on the IC and silicon transistor; Hitachi with National Semiconductor on the semiconductor and IC and manufacturing equipment; Hitachi with Motorola, and Oki Electric with WE on the semiconductor manufacturing equipment; Toko with MOS Technology on the MOS IC and semiconductor; and Sharp with Zilog on the microprocessor.

1978 J: CCD with one chip containing 110,000 elements was developed (SONY).

J: Japanese companies dominated 40% of the 16K-bit DRAM market in the U.S.

J: Hitachi and NEC began establishing production facilities in the U.S.

J: Promotion Law for Specific Machinery and Information Industries (*Tokutei Kikai Joho Sangyo Shinko Rinji Sochi Ho*: *Kijoho*) took over the *Kidenho*.

J: TRI started the second VLSI development project (1978-1980).

J: World's first x-ray lithography system (TRI and Nikon).

1979*J: Fujitsu started establishing production facilities in the U.S.
      J: Government established a research program for developing technologies needed
           for the next generation computer (1979-1983) (*Jisedai Konpyutayo Kihon
           Gijutsu Kaihatsu Hojokin*).
      J: Patent on the Poli-Si self alignment technology (NEC).
      J: Surplus, for the first time, in the balance of IC trade.
     JJ: Joint development: VLSI Cooperative and Japan Electron Optics Laboratory on
           the electron beam lithography technology for one micron width.
      J: Development of the epitaxial wafer mass-production technology for GaAs IC
           (Sumitomo Electric Industries).
      J: Clarion entered the IC market.

1980*J: IC exports to the U.S. became larger than imports for the first time.
      J: Toshiba invested in the U.S.
      J: Share of Japanese semiconductor manufacturing equipment drastically increased
           in the Japanese market.
      J: Suwa Seikosha and Ricoh entered semiconductor business.
     UJ: Analog Devices established a subsidiary in Japan.
     UJ: National Semiconductor established a Japanese subsidiary, National Semicon-
           ductor Japan.

1981*J: Government established two research projects for developing (1) high speed
           computer for scientific research (1981-1989) (*Kagaku Gijutsuyo Kosoku Keisan
           Sisutemu*) and (2) new materials for IC (1981-1990) (*Shin Kino Soshi Kaihatsu*).
      J: TRI started the third VLSI development project (1981-1982).
      U: Intel Japan established a design center in Japan.
      J: Olympus entered the semiconductor market.

1982 J: Japanese companies dominated 70% of the world market in the 64K-bit DRAM
           and 44% in the 16K-bit DRAM.
      U: Motorola established production facilities in Japan.
      J: Government established the fifth generation computer development project
           (1982-1990) (*Daigosedai Konpyuta no Kenkyu Kaihatsu*).
     UJ: Technical agreements: Toshiba and Hitachi to provide the 64K-bit DRAM
           technology to Zilog and Hewlett-Packard respectively.

1983 K: Samsung Electronics succeeded in developing the 64K-bit DRAM IC.
      J: Japan dominated 46% of the memory IC world market.
     UJ: Shortage of ICs in the world market.
      J: Annual average of plant-and-equipment-investment growth rate for the biggest
           five Japanese companies between 1977 and 1982 was 41.6%, while that for U.S.
           companies was 16.8%.
      J: Toshiba and Hitachi dominated 70 to 80% of the 16K-bit SRAM world market.
      J: Nippon Telegraph and Telephone Public Corporation (current NTT) established
           the Atsugi Research Institute to concentrate more on developing forefront
           technology.
     UJ: Joint venture, Asahi Microsystems, established by Asahi Chemical Industry and
           AMI.
      J: Yamaha started selling semiconductors to outside companies.
      J: Fuji Xerox entered semiconductor business.

1984 J: Japan dominated 53% of the world MOS IC market.

     W: World IC market became quite inactive.

     J: Japanese companies dominated more than 90% of the 256K-bit DRAM world market.

     J: Submicron transistor technology (Mitsubishi Electric), the thin film forming technology by the light CVD method, and the stacked LSI process technology were developed (NEC).

     UJ: Technical agreements: Sharp and Wafer Scale Integration on the EPROM IC; and Oki Electric and Excell Microelectronics on the EEPROM IC.

     J: MINEBEA established NMB Semiconductor to enter into semiconductor business.

     UJ: LSI Logic established Nihon LSI Logic in Japan.

     UJ: Micro Power Systems established its subsidiary in Japan.

1985 J: Semiconductor Circuit Design Protection Law (*Handotai Kairo Haichi Ho*) was promulagated.

     UJ: Japan and U.S. agreeed to lower semiconductor tariffs.

     J: Nippon Telegraph and Telephone Public Corporation was privatized to be called NTT.

     J: World first development of GaAs dislocation-free crystal (Sumitomo Electric Industries).

     JJ: Joint ventures: Suwa Seikosha and Orient Watch for entering into assembling-and-testing-semiconductor business; and Kanebo Electronics established by Kanebo and Mitsubishi Electric for assembling and testing bipolar digital ICs.

     UJ: Joint venture: Nihon Semiconductor established by Kawasaki Steel and LSI Logic for the former to enter the custom LSI market.

     EJ: Joint venture: Fuji Electronic Components established by Fuji Electric and Siemens for importing and assembling power MOS FET ICs.

     JJ: Suwa Seikosha and Epson merged to establish Seiko-Epson.

     JJ: Joint development between Matsushita Electronics and Hayashibara Biochemical Laboratories on water-soluble photo polymer.

     UJ: Joint developments: Toshiba and LSI Logic on the gate array; and NTT with TI and Motorola on the semiconductor.

     KU: Technical agreement between Hyundai Electronics and TI.

     EU: Technical agreement between Fuji Electric and Thomson CSF.

     EJ: Technical agreements: Toshiba and Siemens on semiconductor development and manufacturing; Hitachi and Thomson on the semiconductor; and Oki Electric and Thomson Semiconductor for Oki to provide the 64K- and 256K-bit DRAM technologies.

     UJ: Technical agreement between Seiko-Epson and Xlinks on the gate array.

     JJ: Technical agreement between Nippon Steel and Hitachi for the former to enter the semiconductor market.

     UJ: Second-sourcing agreement between Seiko Instruments and Excell Microelectronics on the EEPROM IC and non-volatile RAM IC.

     JJ: Second-sourcing agreement: SONY produced microprocessors for NEC.

     UJ: Manufacturing and sales agreement between Hitachi and Signetics for reciprocally manufacturing and selling the peripheral IC for 16-bit microprocessor.

     J: Tamura Corporation entered the hybrid IC market.

J: Toyota started producing its own semiconductors.

1986 UJ: Joint venture: Mitsubishi Electric, WH, and GE for producing LSIs used by
          power companies.
     JJ: Joint venture and development: 13 companies and the Basic Technology
          Development and Promotion Center established SORTEC for developing
          synchrotron orbital radiation (SOR) equipment.
     JJ: Joint developments: SONY and AMD on the VLSI; Matsushita Electronics and
          NTT on a future generation LSI; and NEC and SONY on the LSI for graphic
          display control.
     UJ: Joint development between Matsushita Electronics and SAE on the micropro-
          cessor.
     UJ: Technical agreements: Seiko-Epson and Lattice Semiconductor on the SRAM
          IC; Toshiba to provide Motorola with the 64K- and 256K-bit DRAM technolo-
          gies; Oki Electric and Catalyst Semiconductor on the EEPROM IC; and Oki
          Electric to obtain the LSI technology on modem from SSI.
     EJ: Technical agreement between Toshiba and Siemens on the standard cell IC.
     JJ: Technical agreement: Asahi Chemical Industry to obtain the LSI technology
          from Hitachi for the former to enter the market.
     UJ: Manufacturing and sales agreement: Zilog to manufacture and sell Hitachi's
          "HD64180."
     J: Kyocera entered into custom LSI design business.
     UJ: Second sourcing agreement between Matsushita Electronics and Intel on the
          microcomputer.

1987 JJ: Joint venture between NEC and Mazda to produce automobile electronic parts.
     UJ: Joint venture: Tohoku Semiconductor established by Toshiba and Motorola.
     UU: Joint development: TI Japan and LSI Japan on the ASIC.
     UJ: Joint developments: NMB Semiconductor and National Semiconductor on the
          SRAM and EPROM IC; and SONY and AMD on the 256K-bit SRAM IC.
     UJ: OEM agreements: Mitsubishi Electric to supply 256K-bit DRAM ICs to
          National Semiconductor; SONY to supply AD converters to GE; Sharp and Fuji
          Electric to supply SRAM ICs to MOS Electronics; and Seiko-Epson, Hitachi
          and Mitsubishi Electric with VLSI Technology.
     UJ: Technical agreements: Oki Electric to obtain the new gate array technology
          from iLSi; Yamaha and Programmable Memory Technology; and New Japan
          Radio to provide the IC technology to Raytheon.
     UJ: Secondary sourcing agreement between NEC and Standard Microsystems on
          the floppy disk control LSI.
     J: Seiko Instruments started the OEM production of ICs.
     J: Tokairika entered the automobile custom IC market.
     J: Nissan started producing its own semiconductors.

1988 UJ: Joint developments: SONY and TI Japan on the high performance digital filter
          LSI for digital audio system; Matsushita Electronics and Intel on the 16M-bit
          DRAM processing technology; Sanyo and VLSI Technology on the one-chip 32-
          bit RISC microcomputer; and Hitachi and TI on the 16M-bit DRAM technol-
          ogy.
     UJ: Technical agreements: Toshiba to provide the 1M-bit DRAM technology to

Motorola; and Hitachi and VLSI Technology on the ASIC.

JJ: Technical agreement: SONY to provide the MCZ technology to Shinetsu Handotai.

EJ: OEM agreement: SGS Thomson Microelectronics Italy to assemble semiconductors for Oki Electric.

UJ: OEM agreement: Mitsubishi to assemble ICs for National Semiconductor.

JJ: OEM agreements: Mitsubishi and Kanebo to assemble high speed transistor-transistor logic ICs for National Semiconductor; and Kanebo Electronics to assemble 1M-bit DRAM ICs for Mitsubishi Electric.

EJ: Mutual supply agreement between Matsushita Electric and Philips on electronics parts.

U: Applied Materials moved its headquarter functions from the U.S. to Tokyo.

K: Samsung Electronics, Hyundai Electronics, and Gold Star Electron made a large investment for 1M-bit DRAM production.

J: Asahikasei Microsystems entered into the production of high speed EPROM IC with technology provided by ICT.

J: Toyota announced to produce five types of electronics parts by itself.

1989 J: Development of the multi-layer technology for analog processor.

UJ: Joint venture: Japan Motorola Microelectronics established by Motorola Japan and Toshiba Device for semiconductor sales.

JJ: Joint venture: VM Technology established by ASCII and Mitsui Trading Co. to enter into the semiconductor market.

JJ: Joint development: 16 companies including NHK, Toshiba, NEC, etc. on the first generation IC for High Density Television (HDTV).

UJE: Joint development: NEC, Siemens, and MIPS Computer on the next-generation microprocessor.

UU: Joint development: AT&T Microelectronics and Xerox on the ASIC for Xerox machines.

UJ: Joint developments: Sharp and SMC on the MOS technology; Toshiba and ICS on gate array testing and designing systems; Toray and OSI on IC test systems for the former to enter into semiconductor manufacturing equipment business; and Fujitsu and TI Japan to make their ASIC software compatible.

UJ: OEM agreements: Motorola to produce CMOS logic ICs for Sanyo; and LSI Logic to produce 256K-bit SRAM ICs for Sharp.

UJ: Mutual supply agreements: Hitachi and National Semiconductor on logic ICs; and Hitachi and TI on 64K-bit SRAM ICs.

KJ: Technical agreement: Hitachi to provide the 1M-bit DRAM technology to Gold Star Electron.

UJ: Technical agreements: MIPS Computer to provide SONY with the design technology of 32-bit RISC microprocessor; and Fujitsu to provide Vitesse Semiconductor the GaAs circuit technology.

EJ: Technical agreement: Oki Electric to provide the 1M-bit DRAM technology to SGS Thomson.

J: Nippon Steel decided to establish VLSI sample production lines and a research facility.

UJ: Nippon Steel entered into ASIC designing business, and contracted VLSI Technology for production.

J: Ishikawajima-Harima Heavy Industries announced to enter into semiconductor test system business.

J: NKK announced entering into VLSI business, producing 4M-bit DRAM ICs.

U: IBM established the Semiconductor Research Center in New York and gathered world-wide semiconductor research facilities in one place, aiming for future devices with SOR technology.

U: TI Japan established a semiconductor basic research center in Tsukuba.

U: LSI Logic shifted the center of semiconductor production from California to Japan.

UJ: TDK entered the semiconductor market by acquiring Silicon Systems in the U.S. and exporting products to Japan.

1990 UJ: Second generation IC for HDTV was developed by several alliances: (1) Toshiba-Motorola; (2) Sanyo-LSI Logic; (3) SONY TI Japan-Hitachi-Fujitsu; and (4) NEC-Mitsubishi Electric–Matsushita Electronics.

J: Toshiba opened its semiconductor training facilities to cooperating companies.

J: World's first low-temperature BiCMOS IC (NEC).

K: Samsung Electronics started the sample production of 16M-bit DRAM IC.

K: Samsung Electronics succeeded in developing the 4M-bit Mask ROM IC.

K: Samsung Electronics started the sample production of 4M-bit DRAM IC.

UJ: Joint developments: Mitsubishi and AT&T Microelectronics on the GaAs mass production technology; and SONY and AMD on the 1M-bit SRAM IC.

JJ: Joint developments: Toshiba and Nikon on the 16M-bit DRAM processing technology; and Fujitsu, Sharp, and Kyodo Electronics to develop the linear IC for industrial use.

JJ: Joint venture between Mitsubishi Electric and Asahi Glass to enter the liquid crystal display (LCD) market.

UJ: Joint venture: KTI Semiconductor established by KOBELCO and TI for the former to enter into the semiconductor market.

UJ: OEM agreements: NMB Semiconductor to produce DRAM ICs for Intel; NEC to produce 256K-bit SRAM ICs for National Semiconductor; Hitachi to produce cooperatively developed SRAM ICs for VTI; and NEC to produce 4-bit microprocessors for AT&T Microelectronics.

EJ: OEM agreement: AEG to produce semiconductors for Mitsubishi.

UJ: Technical agreements: Tohoku Semiconductor (a Toshiba-Motorola joint venture) to produce Motorola's 32-bit microprocessors and Toshiba's 4M-bit DRAM ICs; Toshiba and Motorola on the ASIC; Toshiba and MIPS Computer to produce and sell MIPS-designed RISC microprocessors under the MIPS brand; Toshiba and IBM Japan to jointly produce color liquid crystal displays; Toshiba to provide Harris Corporation the ASIC technology for the latter to establish an ASIC design center in Tokyo; Mitsubishi Electric to provide AT&T Microelectronics the SRAM technology to be produced and sold in the U.S.; Mitsubishi Electric to provide AT&T Microelectronics the GaAs IC technology and the latter to produce and sell in the U.S.; Intel to provide Matsushita Electronics the microprocessor technology to produce Intel's microprocessors for sale in Japan; NEC to provide AT&T Microelectronics the gate array technology; AT&T Microelectronics to provide NEC a CAD system for gate array and standard cell design; and Hewlett-Packard to provide Hitachi the designing technology of RISC microprocessor for the latter to enter the market.

EJ: Technical agreement: Oki Electric to provide SGS Thomson the 4M-bit DRAM technology to jointly produce and sell in Europe.

KJ: Technical agreement: Hitachi to provide Gold Star Electron the 4M-bit DRAM processing technology.

J: NKK established the Electronic Device Research Laboratory to acquire the ASIC development and production technologies.

1991 UJ: Joint developments: Kawasaki Steel and Ten Sleep to develop application specific standard products (ASSPs); Sharp and Applied Materials on the liquid crystal display technology; AT&T Microelectronics and NEC to develop the 0.35 micron CMOS IC process technology; Hitachi and TI to develop the 64M-bit DRAM technology; and Tohoku University, Chisso, and Bellcore to develop the electron beam photo-resist technology for 1 giga-bit ULSI.

UJ: Technical agreements: Matsushita Electronics and Weitek on the high speed LSI; and Hitachi to provide NMB Semiconductor the 4M-bit DRAM technology.

UJ: OEM agreement: Signetics to produce wafer and assemble ICs for Matsushita Electronics.

JJ: OEM agreement: NMBS to produce ICs for Hitachi.

J: NTT developed the SOR technology with 0.2 micron width.

J: Fujitsu developed a computer with 64-bit SRAM HEMT LSIs for the first time in the world.

1992 UJ: Joint developments: Toshiba and ITD on the RISC-type microprocessor; and NEC and AT&T Microelectronics on the 4M-bit SRAM IC.

UJ: Technical agreements: Fujitsu to provide Samsung access to all of its semiconductor technology; Matsushita Electronics and Symetrix on ferro-electric materials; Sharp to provide Intel the flash memory technology; and Toshiba and Intel to expand free cross-licensing on technology.

EJ: OEM agreement: AEG to produce bipolar ICs for Mitsubishi Electric.

J: Hitachi established a research facility for developing the 1 giga-bit DRAM IC.

UJ: SONY, NEC, Hitachi, Fujitsu, Matsushita Electronics, and Motorola to participate in the government's next generation project, "Quantum Functional Device Development Research," organized by the Research and Development Association for Future Electron Devices.

Note: J: Japan; U: U.S.; E: Europe; K: Korea; W: world; *: data on technical agreements not available.
Source: Denpa Shinbunsha, *Denshi kogyo nenkan* (Yearbook of electronics industry), (Tokyo: Denpa Shinbunsha, 1964-1992).

Yasuzo Nakagawa, *Nihon no handotai kaihatsu* (Development of semiconductors in Japan), (Tokyo: Diamond, 1985).

Press Journal, *Nihon handotai nenkan* (Japan semiconductor yearbook), (Tokyo: Press Journal, 1985-1992).

Sangyo Times, *Handotai sangyo keikaku soran* (Comprehensive list of semiconductor industry planning), (Tokyo:Sangyo Times, 1985-1991).

Nihon Keizai Shinbun, "*Motorola, Bei de DRAM seisan kaishi –Toshiba no gijutsu kyoyo* (Motorola produces in the U.S. with Toshiba technology)," April 20, 1988, Morning Version, 8.

_____, "*Matsushita, Intel, 16M DRAM no kako gijutsu kyodo kaihatsu e* (Matsushita Electronics and Intel jointly develop 16M DRAM)." Dec. 30, 1988, Morning Version, 6.

_____, "*Fujitsu, garihiso handotai de teikei – Bei Vitesse-sha to* (Fujitsu to cooperate with Vitesse on GaAs semiconductor," November 9, 1989, Evening Version, 2.

_____, "*Nichiden, Bei NS ni OEM kyokyu – 256 SRAM, tsuki ni 10-20 manko* (NEC to OEM-supply 256 SRAM to NS, 100-200 thousand per month)." May 29, 1990, Morning Version, 11.

_____, *"Fujitsu, HEMT LSI konpyuta ni sekai de hajimete tosai* (Fujitsu to use HEMT LSI in computer, the first time in the world)," February 13, 1991, Morning Version, 10.

_____, *"Toshiba, Bei IDT to teikei* (Toshiba contracts technical agreement with IDT of the U.S.)," February 20, 1991, 13.

_____, *"Sansei Denshi Fujitsu e 40 okuen* (Samsung pays Fujitsu 4 billion yen)," April 22, 1991, 1.

_____, *"Matsushita Denshi, Bei handotai kaisha to teikei –LSI no seizo gijutsu de* (Matsushita Electronics to provide LSI technology to U.S. semiconductor company)," May 16, 1991, Morning Version, 10.

_____, *"4M DRAM, Minebea kei NMBS shien – Hitachi ga seisan itaku* (Hitachi supports NMBS with 4M DRAM OEM production)," June 20, 1991, Morning Version, 1.

_____, *"Matsushita Denshi, Philips Bei kogaisha to handotai de kyoryoku kyoka* (Matsushita Electronics strengthens ties with Philips' U.S. subsidiary on semiconductors," July 23, 1991, 13.

_____, *"Sharp, ekisho hyoji sochi o Beisha to kaihatsu e* (Sharp and a U.S. firm to develop liquid crystal display equipment)," November 12, 1991, Morning Version, 11.

_____, *"64M DRAM, Hitachi, TI to teikei – R&D kosuto o buntan* (Hitachi and TI cooperate for 64M DRAM development to share costs," November 21, 1991, Morning Version, 13.

_____, *"Matsushita Group, Philips o handotai shien – Bei kokaisha ni seisan itaku* (To support Philips on semiconductors, Matsushita Group contracts OEM production with Philips' U.S. subsidiary," December 10, 1991, Morning Version, 11.

_____, *"Hitachi 1 giga DRAM kaihatsu chakushu, chuoken ni shin kenkyu shisetsu* (Hitachi establishes new facility for developing 1-giga DRAM)," January 29, 1992, 11.

_____, *"Motorola, SONY nado, 6 sha no sanka kettei – ryoshika kino soshi no kenkyu kaihatsu* (Six companies including Motorola and SONY participate in developing quantum function devices," January 30, 1992, Morning Version, 1.

Nikkan Kogyo Shinbun, *"Nichiden, Bei Maikuroere to teikei – handotai seihin no gijutsu ya seisan de* (NEC to cooperate with AT&T Microelectronics in semiconductor technology and production," March 8, 1990.

_____, *"NEC, teion BiCMOS no sado ni sekai de hajimete seiko* (NEC successful in low temperature BiCMOS, the first time in the world)," December 21, 1990, 11.

_____, *"Nichiden, AT&T to kyodo de jijisedai no LSI kaihatsu e* (NEC to develop two-generation-advanced LSI with AT&T cooperatively)," April 23, 1991, 11.

_____, *"HDTVyo IC, kasoku suru kyodo kaihatsu* (HDTV-IC, speeding up cooperative research)," November 6, 1991, 9.

_____, *"Toshiba, U.S. Intel, handotai gijutsu ni kansuru musho kurosu raisensu keiyaku no taisho o kakudai* (Toshiba and U.S. INTEL to expand free cross-licensing agreements)," January 30, 1992, 9.

Nikkei Sangyo Shinbun, *"Sharp, Bei Intel furasshu memori de teikei* (Sharp to provide Intel the flash memory technology)," February 6, 1992.

_____, *"Kyo yudentai usumaku no seizo gijutsu, Matsushita Denshi, Beisha to teikei* (Matsushita Electronics to have technical agreement with U.S. company on ferro electric materials), March 19, 1992, 7.

Appendix Table 2.4 Development of Organizational Goals, Structure, and Strategies for Semiconductor Production in Company A

	Organizational Goals	Organizational Structure	Strategies to Attain Goals
1951	To acquire new technologies from foreign companies through licensing	Beginning of research on the transistor in the Central Research Laboratory (prior to this year there was an informal research group)	
1952			A technical agreement with RCA
1953			
1954			A patent agreement with WE
1955			
1956	To establish technological base for transistor production	The Transistor Section in the Central Research Laboratory established	
1957			
1958	To establish mass production facilities	The Transistor Laboratory made independent from the Central Research Laboratory	
1959		The Transistor Laboratory renamed as the AA Works in the Communication Equipment Division	1st Research Report Presentation
1960			A training school established within the AA Works 2nd Research Report Presentation
1961	To rationalize production	The AA Works renamed as the Electronics Tube Division and made independent from the Communication Machinery Division	QC promotion activities Suggestions for reducing production costs Production processes automated
1962	To concentrate on Si transistor development with strong emphasis on technological self-reliance		Production cost cutting activities Production processes automated
1963	To upgrade technological capabilities to catch up with U.S. companies Product specialization	The AB Electronics Co. Ltd. established for the production of parts	
1964	To improve quality and reduce production costs for winning competition in the over-supplied market	The AC Branch established for producing diodes	

	Organizational Goals	Organizational Structure	Strategies to Attain Goals
1965	To develop ICs and products using ICs (calculators, watches, televisions, etc.)	The AD Electronics Co. Ltd. established for the production of thermistor; The Temporary IC Development Dept. jointly established by the Central Research Laboratory and the AA Works	Reducing rejection activities (*furyo teigen undo*) Licensing with RCA
1966		The Mechanical Engineering Research Laboratory, established directly under the President	
1967	To mass produce ICs	The AE Branch for Ge Transistor established; The IC Dept. established in the AA Works to concentrate on IC production; A research branch of the Central Research Laboratory established within the AA Works to develop new products and process technologies	
1968	To rationalize administration with computers and to automate production processes	The Quality Control Department established	
1969	To develop transistor products with self-developed processing technology; To increase international competitiveness by improving quality and quantity in IC production	The Semiconductor and IC Division and the Process Technology Development Department established; The AC and AE Branches made independent within the new Semiconductor and IC Division	Lecture series on IC technologies to workers
1970		The AF Works established for the production of Si transistors, and the AA Works made to concentrate on IC production	A training factory established; Zero-defect activities
1971		The Crystal Section moved to the AC Works, and the Ge Diode Section to the AE Works; The Production Engineering Research Laboratory established directly under the president	
1972	To increase LSI development capability	The Electronics Communication Development Dept. established to develop ICs and LSIs for computers and communication equipment; The AG Semiconductor Malaysia established	

	Organizational Goals	Organizational Structure	Strategies to Attain Goals
1973	To increase the number of patents; to decrease dead stocks, required time for new technological development and customer claims; to simplify and shorten production processes; and to reduce the number of defective goods, paperwork, time for meetings, and accidents	The Systems Development Laboratory established directly under the president	Management Improvement "MI Challenge 50 Activities"
1974			Management Improvement "MI Fresh Activities;" Doubling VA (Value Analysis) activities
1975	To shift focus from consumer goods to industrial goods	The Device Development Center established by merging the Computer and Communication IC Development and the Experiment Departments; The CAD Center established; The AC and AE Branches brought under the AA Works and AF Works, respectively	"MMM Activities" (Challenge-to-zero defect activities by reducing excess (*muri*), waste (*muda*), and unevenness (*mura*)
1976	To develop indigenous micro-processors in order to cope with trade frictions; To cultivate the system designing capability		"MMM II Activities"
1977	To develop new products and engage in mass production	The Process Technologies Development Department and the Designing and Development Department moved into the AA Works; Independent departments for memories and microcomputers established	"MMM Dash 20 Activities"
1978		AH Semiconductor (America) Inc. established	"MMM Scrum 80 Activities"
1979			
1980		The AI Microcomputer Engineering Ltd. established as an independent company to engage in the development of microprocessors; AJ Semiconductor (Europe) GmbH. established	"MMM Wandering Samurai 200 Up Activities"
1981			
1982			"MMM Power-Up Activities"
1983			
1984		The AK VLSI Engineering Co. established as an independent organization to engage in the development of VLSI technology	"MMM Challenge-Fresh Activities"

Organizational Goals	Organizational Structure	Strategies to Attain Goals
1985	The Advanced Research Laboratory established to promote basic research	"MMM Step-Up-Musashi Activities"
1986		"MMM Fresh-Attack Activities"

Note: A, AA, AB, AC, AD, AE, AF, AG, AH, AI, AJ, and AK are used to maintain the anonymity of the
        interviewed company. Dept.: Department
Source: AA Factory Twenty Year History; and interviews with four top-level executives

Appendix Table 2.5 Development of Organizational Goals, Structure, and Strategies for Semiconductor Production in Company B

	Organizational Goals	Organizational Structure	Strategies to Attain Goals
1949		A few researchers started conducting research within a factory without any research funding and any official approval (the first among other companies); The Central Research Institute closed due to financial difficulty	
1950		Research on the transistor started officially in the company	
1951			
1952			
1953			
1954		The Central Research Institute reopened to concentrate on both basic and applied research; Research on a point contact diode and alloying and grown transistors; The Semiconductor Sub-section, established to prepare for mass production within the Process Technology Section of the Electron Tube Industry Division in the Electronic Components Joint Division	
1955	To self-develop transistor production technology		Research team failed in building mass production capability without foreign licensing
1956		The Semiconductor Sub-section, developed into the Semiconductor Development Dept. within the Electron Tube Industry Division (*Denshikan Kogyobu*); A few researchers from TRI and three or four researchers from ETL moved to the new department	A training center established within a factory
1957	To search for foreign patent and technology contract partners		
1958	To shift transistor production from computer-and-communication equipment to commercial-use equipment in order to generate scale merit	A patent agreement contracted with RCA and a technical agreement with GE; The production and quality control system introduced The Semiconductor Development Dept. changed to the Semiconductor Dept. in the Electron Tube Industry Division (*Kogyobu*)	

	Organizational Goals	Organizational Structure	Strategies to Attain Goals
1960	A top executive emphasized the importance of developing communication equipment and computers with ICs	The Semiconductor Dept. changed to the Semiconductor Industry Division and made independent from the Electron Tube Industry Division; Within the factory the Development Department engaged in research on new products; Five research institutes established: (1) basic technology; (2) communication technology; (3) electron device; (4) atomic physics; and (5) production technology	A technical school established
1961		A task force for the production of IC established; Under a new division (*jigyobu*) system, the Semiconductor Industry Division belonged to the Electron Components Group	
1962		The Central Research Institute started basic research on the GaAs IC	
1963			
1964			
1965	Management reform for internationalization by (1) establishing a top management group, (2) strengthening the division system, (3) reorganizing a more effective R&D system, (4) reorganizing five research institutes (RI) into three (the Basic RI, the Communication RI, and the Electron Device RI), (5) introducing a project group system, and (6) intensifying zero-defect activities; to automate the production system; to develop products with ICs and LSIs		Management improvement activities: A zero-defect (ZD) committee established to promote ZD activities
1966		Three IC-related R&D centers established within the Semiconductor Industry Division to enhance R&D activities in the areas closely related to products and production: (1) the Integrated Circuit Design Center; (2) the Semiconductor & IC Production Center; and (3) the Thin Film Semiconductor Production Center	

	Organizational Goals	Organizational Structure	Strategies to Attain Goals
1967		The Electronic Components Group was split into the Electron Device Group and the Semiconductor and IC Group; The three IC related R&D centers were unified into the Semiconductor and IC Group	The Reliability Control Dept. established directly under the President
1968		One section of the Central Research Institute became independent as the Microelectronics Research Institute	
1969			
1970			
1971			"Spark 10 Activities" in order to reduce production costs by 10%
1972		The Electron Device Group and the Semiconductor and IC Group were recombined and named as the Electron Device Group; The Electron Device Group Reliability and Quality Control Center and the Electron Group Planning Office established and controlled directly by the head of the Group	"Quality Strategy Activities" to improve the quality of management, products, services, the environment, public relations, achievements, and company image; The establishment of a system to encourage all employees to participate in R&D
1973	To automate production for ICs and LSIs and administration, using computers		
1974			"Quality Strategy Activities"
1975		The VLSI R&D Center in the Electron Device Group established	A training center established
1976			
1977	To complete automation in IC and LSI production		
1978			
1979			A technical training center established
1980			
1981			

Organizational Goals	Organizational Structure	Strategies to Attain Goals
1982	The IC and the Semiconductor Divisions renamed as the First and the Second LSI Divisions	
1983		
1984	Three step R&D system established: (1)Basic research in the Basic Research Institute and the Microelectronics Research Institute; (2)Applied research for technology development in the VLSI R&D Center and the VLSI & CAD Technique Development Center; and applied research for promoting commercial product development in the Memory Technology Center, the Microcomputer Technology Center and the Custom LSI Technology Center; and (3)Immediate technological and commercial research in the Process Development Dept., the Automation Promotion Dept., and the New Products Dept. in the factory	
1985	The Electron Device Group Patent Technology Center and the Electron Device Group Analysis Technology Center added as a part of the support system under the direct control of the group head; The Electronic Components Subgroup added to facilitate the coordination and allocation of component supplies	
1986		

Note: Dept.: Department

Source: "Recent Ten-Year History of Company B" and "Seventy-Year History of Company B" published
by Company B; and Interviews with six top executives of Company B

Appendix Table 5.1 Flexible-synergy and Market-like Effects by Type of Interaction in R&D

BENEFITS	TYPE OF INTERACTION			
	SPOT INTERACTION (N = 6)	HORIZONTAL INTERACTION (N = 6)	VERTICAL INTERACTION (N = 7)	NTERNALIZED INTERACTION (N = 6)
A. Flexible-synergy Effects				
1. From Long-term Mutual Accumulation				
Acquiring Technological Information*	2.00	3.83	4.33	4.50
Acquiring R&D Information*	2.00	3.67	4.00	4.50
Obtaining Joint R&D Projects, owing to Long-term Mutual Experience*	2.00	2.83	3.67	4.00
Enhancing Trust and Cooperation*	1.50	4.00	4.00	4.00
Reducing R&D Risk*	2.50	3.50	4.00	3.50
2. From Satisfying Mutual Expectations				
Coordinating R&D*	2.00	3.17	3.67	4.00
Sharing Human Resources and Equipment*	2.50	3.00	4.00	4.17
Keeping Corporate Secrets*	2.00	3.00	2.67	4.33
3. From Future-oriented Risk-taking Activities				
Stimulating Creativity*	2.00	3.33	3.33	4.00
Developing Future-oriented Risk-taking Projects*	2.00	3.50	3.33	4.00
B. Market-like Effects				
Maintaining Strict Standards and Tough Relationship*	2.00	3.00	2.67	4.33
C. Common Goals and Interests				
	-Quick purchase of minor technology	-Technology sharing -Cost sharing -Product market leadership -Quality -Product market sharing -Reduced time for innovation	-Technology sharing -Cost sharing -Mutual growth	-Product market leadership -Developing advanced devices

Note: * 1.0 means very weak, and 5.0 means very strong. Scores are categorized as follows: very weak (mean value = 1.0 - 1.4); weak (1.5 - 1.9); weak + (2.0 - 2.4); medium (2.5 - 2.9); medium + (3.0 - 3.4); strong (3.5 - 3.9); strong + (4.0 - 4.4); and very strong (4.5 - 5.0).

Bibliography

Abegglen, James C. 1984. Strategy of Japanese business. Cambridge, Mass.: Ballinger
    Publishing Co.
Abegglen, James C., and George Stalk, Jr. 1985. *Kaisha* (Company). Tokyo: Charles E.
    Tuttle Co.
Aida, Yutaka. 1991. *Denshi rikkoku Nihon no jijoden* (Autobiography of electronic
    country Japan) I. Tokyo: NHK.
———. 1992. *Denshi rikkoku Nihon no jijoden* (Autobiography of electronic country
    Japan) III. Tokyo: NHK.
Allen, George. C. 1981. A short economic history of modern Japan. London: Veidenfeld
    and Nicolson.
Anchordoguy, Marie Christine. 1986. The state and the market: Industrial policy toward
    Japan's computer industry. Ph.D. Thesis. University of California-Berkeley.
Aoki, Masahiko. 1984. The cooperative game theory of the firm. London: Clarenden
    Press of Oxford University Press.
———. 1988. Information, incentives, and bargaining in the Japanese economy. Cam-
    bridge: Cambridge University Press.
Asanuma, Banri. 1985. "The organization of parts purchases in the Japanese automo-
    tive industry." Japanese Economic Studies (Summer): 32–53.
Asada, Akira. 1984. *Kozo to chikara* (Structure and Power). Tokyo: Keiso Shobo.
Baranson, Jack. 1981. The Japanese challenge to U.S. industry. Lexington, Mass.:
    Lexington Books.
Befu, Harumi. 1986. "The social and cultural background of child development in
    Japan and the United States." In Child Development and Education in Japan,
    edited by H. Stevenson, H. Azuma and K. Hakuta. New York: W.H. Freeman.
Bellah, Robert N. 1957. Tokugawa religion: The cultural roots of modern Japan. N.Y.:
    Free Press.
Benedict, Ruth. 1969. The chrysanthemum and the sword: Patterns of Japanese culture.
    Tokyo: Charles E. Tuttle Co.
Calder, Kent E. 1988. Crisis and compensation: Public policy & political stability in
    Japan, 1949–86. Princeton, New Jersey: Princeton University Press.
Campbell, John L., and Leon N. Lindberg. 1990. "The evolution of economic gover-
    nance." A paper presented at the Annual Meeting of the American Sociological
    Association in August 1990 in Washington, D.C.

Chamberlain, Neil, and James Kuhn. 1965. Collective bargaining. New York: McGraw Hill.

Chandler, Alfred. 1977. The visible hand: The managerial revolution in American business. Cambridge: Harvard University Press.

Chunichisha. 1997. *Denshi buhin nenkan 1996/97* (Annual of electronic devices and components). Tokyo: Chunichisha.

Cole, Robert. 1979. Work, mobility, and participation. Berkeley: University of California Press.

Cole, Robert, and Taizo Yakushiji. 1984. The American and Japanese auto industries in transition. Report of the Joint U.S.-Japan Automotive Study, Center for Japanese Studies, The University of Michigan, Ann Arbor.

Contractor, Farok J., and Peter Lorange. 1988. Cooperative strategies in international business. Lexington, Mass.: Lexington Books.

Crawford, C. Merle. 1983. New products management. Homewood, Ill.: Richard D. Irwin, Inc.

Cremer, Jacques. 1990. "Common knowledge and the co-ordination of economic activities." In The firm as a nexus of treaties, edited by Masahiko Aoki, Bo Gustafsson, and Oliver E. Williamson, 53–76. London: Sage Publications.

Denpa Shinbunsha. 1962–1997. *Denshi kogyo nenkan* (Yearbook of electronics industry). Tokyo: Denpa Shinbunsha.

Department of Defense, U.S. Government. 1987. "Defense semiconductor dependency." Quoted in High Technology Strategy Study Group. 1988. *Beikoku no gijutsu senryaku* (Technological strategy of the U.S.). Tokyo: Nikkei Science.

Dore, Ronald. 1986. Flexible rigidities. Stanford: Stanford University Press.

———. 1987. Taking Japan seriously. Stanford: Stanford University Press.

Dyer, Davis, Malcolm S. Slater, and Alan M. Webber. 1987. Changing alliances. Boston, Mass.: Harvard Business School Press.

Electronic-World-News. 1991. "Convex goes with GaAs for C3800." Electronic-World-News, 20 May.

Electronics-Times. 1991. "Inmos delivers first customized transputer." Electronics-Times, 14 November.

Finan, William F., and Jeffrey Frey. 1994. *Nihon no gijutsu ga abunai* (Japan's crisis in electronics: Failure of the vision). Translated by Toshiaki Ikuta and Yukiko Kurihara. Tokyo: Nihon Keizai Shinbunsha.

Fransman, Martin. 1990. The market and beyond: Cooperation and competition in information technology development in the Japanese system. Cambridge: Cambridge University Press.

Fruin, Mark. 1992. The Japanese enterprise system: Competitive strategies and cooperative structures. Oxford: Clarendon Press.

Fukayama, Masamitsu; Saburo Yamashita; and Hiroshi Sanuki. 1986. *Rinkyoshin de kyoiku wa donaru* (How can the Conference on Culture and Education change Japanese education?). Tokyo: Rodo Junposha.

Futatsugi, Yusaku. 1990. "What share cross-holdings mean for corporate management." Economic Eye 11 (1, Spring): 17–19.

Gerlach, Michael. 1989. "*Keiretsu* organization in the Japanese economy: Analysis and trade implications." In Politics and Productivity, edited by Chalmers Johnson, Laura D'Andrea Tyson, and John Zysman, 141–74. New York: Ballinger Publishing Co.

Granovetter, Mark. 1985. "Economic action and social structure: The problem of embeddedness." American Journal of Sociology 91 (3): 481–510.

Gregory, Gene. 1985. Japanese electronics technology: Enterprise and innovation. N.Y.: John Wiley & Sons.

Hall, Peter. 1986. Governing the Economy. New York: Oxford University Press.

Hamaguchi, Eshun. 1977. *Kanjinshugi no shakai Nihon* (Japanese society based on interpersonalism). Tokyo: Toyo Keizai Shinposha.

Hazama, Hiroshi. 1984. *Nihon romu kanri shi kenkyu: Keiei kazokushugi no keisei to tenkai* (History of personnel management: Formation and development of familism). Tokyo: Ochanomizu Shobo.

Hollingsworth, J. Rogers. 1991. "The logic of coordinating American manufacturing sectors." In Governance of the American economy, edited by John L. Campbell, J. Rogers Hollingsworth, and Leon N. Lindberg, 35–73. Cambridge: Cambridge University Press.

Hollingsworth, J Rogers, and Robert Boyer. 1997. Contemporary Capitalism: the Embeddedness of Institutions. N.Y.: Cambridge University Press.

Howard, Robert. 1990. "Can small business help countries to compete?" Harvard Business Review (November-December): 88–103.

Howell, Thomas R., William A. Noellert, Janet H. MacLaughlin, and Alan W. Wolff. 1988. The microelectronics race: The impact of government policy on international competition. Boulder: Westview Press.

Ikeda, Masataka. 1990. "*Senbetsu/shuyakuka de buhin gyokai mo saihen e* (Reorganization of automobile parts industry by screening and concentration)." *Ekonomisuto* (February 13): 62–65.

Imai, Ken'ichi. 1990. *Joho nettowaku shakai no tenkai* (Information network society). Tokyo: Chikuma Shobo.

Imai, Ken'ichi, ed. 1986. "*Inobeshon to kigyo no senryaku/soshiki* (Innovation and corporate organizational strategy." In *Inobeshon to soshiki* (Innovation and organization). 3–32. Tokyo: Toyo Keizai Shinposha.

Imai, Ken'ichi, and Ikuyo Kaneko. 1988. *Nettowaku soshikiron* (Network organizations). Tokyo: Iwanami Shoten.

Imai, Masaaki. 1986. *Kaizen* (Improvements). New York: Random House Business Division.

Inoue, Munemichi. 1985. "Competition and cooperation among Japanese corporations." In The management challenge: Japanese view, edited by Lester C. Thurow, 139–59. Cambridge, Massachusetts: M.I.T. Press.

INSEC. 1988. "*Handotai bunya ni okeru Nihon no shoshukan chosa hokokusho* (Research report on Japanese business practices in the semiconductor industry." Report of the Research Committee, International Semiconductor Cooperation Center Technology Committee. March.

————. 1991. "Difference between Japanese and foreign-based suppliers." Report by The Technology Committee, International Semiconductor Cooperation Center. March.

Inukai, Ichiro. 1981. "Experience in transfer of technology from the West: Lessons from false starts." In Nation-Building and Regional Development, edited by H. Nagamine, 77–98. Nagoya: United Nations Center for Regional Development.

Ishikawa, Kaoru. 1985. What is total quality control? Translated by David J. Lu. Englewood Cliffs, N.J.: Prentice-Hall, Inc.

Iwata, Ryushi. 1977. *Nihonteki keiei no hensei genri* (Principles of Japanese management). Tokyo: Bunshindo.

————. 1982. Japanese management: Its foundation and prospects. Tokyo: Asian Productivity Organization.

————. 1984. *Nihonteki keiei ronso* (Disputes over Japanese management). Tokyo: Nihon Keizai Shinbunsha.

Jankowski, Richard. 1989. "Preference aggregation in firms and corporatist organizations: The enterprise group as a cellular encompassing organization." American Journal of Political Science 33(4): 973–996.

Johnson, Chalmers. 1982. MITI and the Japanese miracle. Stanford, Calif.: Stanford University Press.

Jones, Daniel T. 1985. "The Internationalization of the automobile industry." Journal of General Management 10 (3): 23–44.

Jorde, Thomas M., and David J. Teece. 1989. "Competition and cooperation: Striking the right balance." California Management Review 31 (3): 25–37.

Kawashima, Takenori. 1950. *Nihon shakai no kazokuteki kosei* (Family Structure of Japanese Society). Tokyo: Nihon Hyoronsha.

Keizai Koho Senta. 1995. Japan 1996. Tokyo: Keizai Koho Senta.

Khalil, Elias L. 1995. "Organizations versus institutions." Journal of Institutional and Theoretical Economics 151(3): 445–466.

Kikai Shinko Kyokai Keizai Kenkyusho and Nihon Denshi Buhin Sinraisei Senta. 1980. "*Sangyo yo denshi kiki ni shiyo suru denshi buhin no gijutsuteki sokutei ni kansuru jittai* (Facts about testing electronics parts used for industrial electronic equipment)." February.

Kinzley, W. Dean. 1991. Industrial harmony in modern Japan: The invention of a tradition. New York: Routledge.

Kogut, Bruce, and Udo Zander. 1993. "Knowledge of the firm: The evolutionary theory of the multinational corporation." Journal of International Business 24(4): 625–645.

Kokusai Rengo (United Nations). 1963. *Sekai tokei nenkan* (Statistical yearbook). New York: United Nations.

Langlois, Richard N., and Paul L. Robertson. 1995. Firms, markets, and economic change: A dynamic theory of business institutions. London: Routledge.

Lebra, Takie Sugiyama. 1976. Japanese patterns of behavior. Honolulu: The University Press of Hawaii.

Lindberg, Leon N., John L. Campbell, and J. Rogers Hollingsworth. 1991. "Economic governance and the analysis of structural change in the American economy." In

The governance of the American economy, edited by John L. Campbell, J. Rogers Hollingsworth, and Leon N. Lindberg, 3–34. New York and Cambridge: Cambridge University Press.

Lundvall, Bengt-Ake. 1988. "Innovation as an interactive process: From user-producer interaction to the national system of innovation." In Technical change and economic theory, edited by G. Dosi, C. Freeman, R. Nelson, G. Silverberg and L. Soete, 349–369. London: Pinter Publishers.

Lynn, Leonard. 1984. "Japan adopts a new technology: The roles of government, trading firms, and suppliers." Columbia Journal of World Business (Winter): 39–45.

Markusen, Ann. 1985. Profit cycles, oligopoly, and regional development. Cambridge, Mass.: MIT Press.

Maruyama, Masao. 1961. Nihon no shiso (Japanese thought). Tokyo: Iwanami Shoten.

Masuda Foundation Research Project Team for Japanese Systems. 1992. Japanese systems: An alternative civilization? Yokohama: SEKOTAC.

Minami, Hiroshi. 1955. Gendaijin no shinri (Psychology of contemporary men). Tokyo: Kawade Shinsho.

———. 1980. Nihonjin no ningen kankei jiten (Dictionary of Japanese human relations). Tokyo: Kodansha.

Moos, Felix. 1975. "Acculturation and cultural change: Reflections on the Japanese social structure." In Social structures and economic dynamics in Japan up to 1980, edited by Gianni Fodella, 106–24. Institute of Economic and Social Studies for East Asia, Luigi Bocconi University, Milan, Italy.

Morishima, Michio. 1984. Naze Nihon wa seiko shitaka (Why did Japan succeed?). Tokyo: TBS Britannica.

Murakami, Yasusuke. 1987. "The Japanese model of political economy." In The political economy of Japan: Volume 1 The domestic transformation, edited by Kozo Yamamura and Yasukichi Yasuba. Stanford: Stanford University Press.

Murayama, Motofusa. 1982. "Kazokushugi and shudanshugi management approaches: Source of concept variance in Japanese business settings." In Japanese management: Cultural and environmental considerations, edited by Sang M. Lee and Gary Schwendiman, 171–98. New York: Praeger Publishing Co.

Nagai, Michio. 1971. "Westernization and Japanization: The Early Meiji transformation of education." In Tradition and modernization in Japanese culture, edited by Donald H. Shirely, 35–76. Princeton, New Jersey: Princeton University Press.

Nakagawa, Yasuzo. 1985. Nihon no handotai kaihatsu (Development of semiconductors in Japan). Tokyo: Diamond.

———. 1989. Toshiba no handotai jigyo senryaku (Semiconductor business strategy of Toshiba). Tokyo: Diamond.

Nakamura, Hajime. 1968. Toyojin no shii hoho (Ways of thinking of Orientals). Tokyo: Shunjusha.

Nakamura, Sei. 1989. Chusho kigyo to dai kigyo (Small- and medium-sized enterprises and large enterprises). Tokyo: Toyo Keizai Shinposha.

Nakane, Chie. 1970. Japanese society. Berkeley: University of California Press.

Nakatani, Iwao. 1984. "The economic role of financial corporate groupings." In The economic analysis of the firm, edited by Masahiko Aoki, 227–58. Amsterdam: Elsevier Science Publishers E. V.

Nano, Piko. 1985. *Haiteku saizensen no yomikata* (How to read the forefront of high-technology). Tokyo: Ko Shobo.

Nihon Denshi Kikai Kogyo Kai (Electronics Industries Association of Japan). 1991. '91 IC guidebook. Tokyo: Nihon Denshi Kikai Kogyo Kai.

———. 1994. '94 IC guidebook. Tokyo: Nihon Denshi Kikai Kogyo Kai.

Nihon Denshi Kogyo Shinko Kyokai (Japan Electronic Industry Development Association). 1985. "*Shindenshi zairyo ni kansuru chosa kenkyu hokoku sho XI: Gijutsu joho chosa hokoku* (Research report on new materials in electronics industry: Technical report)." Tokyo: Nihon Denshi Kogyo Shinko Kyokai.

———. 1988. *Denshi kogyo shinko 30 nen no ayumi* (Thirty-year history of electronics industry development). Tokyo: Nihon Denshi Kogyo Shinko Kyokai.

Nihon Keizai Shinbun. 1986. "*IC sangyo futatabi kosei e* (IC industry booming again)." 17 September, 2nd Section.

———. 1988a. "*Motorola, Bei de DRAM seisan kaishi – Toshiba no gijutsu kyoyo* (Motorola produces DRAM in the U.S. with Toshiba technology)." 20 April, morning version, 8.

———. 1988b. "*Matsushita, INTEL, 16M DRAM no kako gijutsu kyodo kaihatsu e* (Matsushita Electronics and INTEL jointly develop 16M DRAM)." 30 December, morning version, 6.

———. 1989. "*Fujitsu, garihiso handotai de teikei – Bei Vitesse-sha to* (Fujitsu to cooperate with Vitesse on GaAs semiconductor." 9 November, evening version, 9.

———. 1990a. "*64M DRAM shisaku seiko* (Success in developing 64M DRAM in experiment)." 8 June.

———. 1990b. "*Nichiden, Bei NS ni OEM kyokyu – 256 SRAM, tsuki ni 10–20 manko* (NEC to OEM-supply 256 SRAM to NS, 100–200 thousand per month)." 29 May, morning version, 11.

———. 1990c. "*Nichibei sangyo kyozon e no michi* (Way for cooperation between Japanese and U.S. Industries)." 8 May, 10.

———. 1991a. "*Fujitsu, HEMT LSI konpyuta ni sekai de hajimete tosai* (Fujitsu to use HEMT LSI in computer, the first time in the world)." 13 February, morning version, 10.

———. 1991b. "*Toshiba, Bei IDT to teikei* (Toshiba contracts technical agreement with IDT of the U.S.)." 20 February, 13.

———. 1991c. "*Sansei Denshi Fujitsu e 40 okuen* (Samsung pays Fujitsu 4 billion yen)." 22 April, 1.

———. 1991d. "*Matsushita Denshi, Bei handotai kaisha to teikei – LSI no seizo gijutsu de* (Matsushita Electronics to provide LSI technology to U.S. semiconductor firm)." 16 May, morning version, 10.

———. 1991e. "*4M DRAM, Minebea kei NMBS shien – Hitachi ga seisan itaku* (Hitachi supports NMBS with 4M DRAM OEM production)." 20 June, morning version, 1.

———. 1991f. "*Matsushita Denshi, Philips Bei kogaisha to handotai de kyoryoku kyoka* (Matsushita Electronics strengthens ties with Philips' U.S. subsidiary on semiconductors)." 23 July, 13.

———. 1991g. "*Sharp, ekisho hyoji sochi o Beisha to kaihatsu e* (Sharp and a U.S. firm to develop liquid crystal display equipment)." 12 November, morning version, 11.

———. 1991h. *"64M DRAM, Hitachi, TI to teikei—R&D kosuto o buntan* (Hitachi and TI cooperate for 64M DRAM development to share costs)." 21 November, morning version, 13.

———. 1991i. *"Matsushita Group, Philips o handotai shien – Bei kogaisha ni seisan itaku* (To support Philips on semiconductors, Matsushita Group contracts OEM production with Philips' U.S. subsidiary)." 10 December, morning version, 11.

———. 1992a. *"Hitachi 1 giga DRAM kaihatsu chakushu, chuoken ni shin kenkyu shisetsu* (Hitachi establishes new facility for developing 1 giga DRAM)." 29 January, 11.

———. 1992b. *"Motorola, SONY nado, 6 sha no sanka kettei – ryoshika kino soshi no kenkyu kaihatsu* (Six firms including Motorola and SONY participate in development of quantum function devices)." 30 January, morning version, 1.

Nikkan Kogyo Shinbun. 1990a. *"Nihondenki, 16M CMOS EPROM o kaihatsu* (NEC develops 16M CMOS EPROM). 14 February.

———. 1990b. *"Nichiden shinkairo kaihatsu* (NEC develops new circuit)." 21 December.

———. 1990c. *"Nichiden, Bei Maikuroere to teikei — handotai seihin no gijutsu ya seisan de* (NEC to cooperate with AT&T Microelectronics in semiconductor technology and production)." 8 March.

———. 1990d. *"NEC, teion BiCMOS no sado ni sekai de hajimete seiko* (NEC successful in low temperature BiCMOS, first time in the world)." 21 December, 11.

———. 1991a. *"Nichiden, AT&T to kyodo de jijisedai no LSI kaihatsu e* (NEC to develop two-generation advanced LSI with AT&T cooperatively)." 23 April, 11.

———. 1991b. *"HDTVyo IC, kasoku suru kyodo kaihatsu* (HDTV-IC, Speeding up cooperative research)." 6 November, 9.

———. 1991c. *"90 nen sekai handotai shea de nihonzei 8 nenburi ni teika* (Lower world semiconductor share in 1990 after 8 years of increase). 11 January, 11.

———. 1992. *"Toshiba, U.S. INTEL, handotai gijutsu ni kansuru musho kurosu raisensu keiyaku no taisho o kakudai* (Toshiba and U.S. INTEL to expand free cross-licensing agreements)." 30 January, 9.

Nikkei Business. 1984a. *"Odoru handotai: Shinario ni kurui wa naika* (Dancing semiconductor industry: No mistake in market prediction?)." 3 September, 22–38.

———. 1984b. *"Gijutsu chinpuka hayaku, doronuma no kato kyoso* (Technology becomes obsolete quickly, forcing firms to overcompete)." 24 September.

Nikkei Sangyo Shinbun. 1991. *"91nen handotai seisangaku, INTEL Bei shui ni, sekai rankingu 4i ni fujo* (91 Semiconductor production, INTEL U.S. no. 1, world no. 4)." 8 January, 6.

———. 1992a. *"Sharp, Bei INTEL furasshu memori de teikei* (Sharp to provide INTEL the flash memory technology)." 6 February.

———, 1992b. *"Kyo yudentai usumaku no seizo gijutsu, Matsushita Denshi, Beisha to teikei* (Matsushita Electronics to have technical agreement with U.S. firm on ferro electric materials)." 19 March, 7.

Nonaka, Ikujiro. 1989. *"Seihin kaihatsu to inobeshon* (Product development and innovation)." In *Nihon no kigyo* (Japanese companies), edited by Ken'ichi Imai and Ryutaro Komiya, 253–71. Tokyo: Tokyo University Press.

North, Douglas. 1989. "Institutions and economic growth: A historical introduction." World Development 17(9): 1319–1332.

―――. 1990. Institutions, institutional change, and economic performance. Cambridge: Cambridge University Press.

―――. 1993. "Institutions and credible commitment." Journal of Institutional and Theoretical Economics 149(1): 1–23.

Oaumann, Edward O., and David Knoke. 1989. "Policy networks of the organizational state: Collective action in the national energy and health domains." In Networks of power: Organizational actors at the national, corporate, and community levels, edited by Robert Perrucci and Harry R. Potter. N.Y.: Aldine de Gruyter.

Odaka, Konosuke, Keinosuke Ono, and Fumihiko Adachi. 1988. The automobile industry in Japan: A study of ancillary firm development. Tokyo: Kinokuniya.

Ohmae, Ken'ichi. 1985. "Managing innovation and new products in key Japanese industries." Research Management 28 (4): 11–18.

Ohno, Minoru, and Fumimaro Kawakatsu. 1983. "Evolution of the semiconductor industry in Japan." VLSI '83, Proceedings of the IFIP TC 10/WG 10.5 International Conference on Very Large Scale Integration, Trondheim, Norway, 16–19 August.

Ohno, Taichi. 1982. "How the Toyota production system was created." Japanese Economic Studies 10 (4, Summer): 83–101.

Okada, Yoshitaka. 1989a. *Nichibei handotai sangyo ni okeru shakai keizai tosei kozo no hikaku* (Comparison of socio-economic coordination structures in Japanese and U.S. semiconductor industries)," In *Kawariyuku nihon no sangyo kozo* (Changing Japanese industrial structure), edited by Akinori Marumo, 52–98. Tokyo: The Japan Times.

―――. 1989b. "Technological development and growth of Japanese integrated circuit firms: An exploratory study." Working Paper at the Center for Japan-U.S. Relations, International University of Japan.

―――. 1990. *Nichibei handotai sangyo ni okeru gabanansu kozo no hikaku I & II* (Comparison of governance structures in Japanese and U.S. semiconductor industries)." *Sekai Keizai Hyoron* (Journal of World Economic Review) (March): 40–53 and (April): 59–65.

―――. 1993. "Institutional arrangements and Japanese competitive-*cum*-cooperative system of production." A paper presented at a conference on the Comparative Market Economies Project held at St. John's College, Cambridge University, September 23–26.

―――, ed. 1999. Japan's industrial technology development: Role of cooperative learning and institutions. Tokyo: Springer-Verlag.

Okada, Y., T. Shishido, T. Hayashi, I Inukai, S. Kimura, H. Uchida, and H. Tada. 1994. "Japan." In Technological independence: The Asian experience, edited by Saneh Chamarik and Susantha Goonatilake, 294–352. Tokyo: United Nations University Press.

Okimoto, Daniel I. 1989. Between MITI and the market: Japanese industrial policy for high technology. Stanford: Stanford University Press.

Otsuka, Hisao. 1973. *Otsuka Hisao zenshu* (Otsuka Hisao series). Vol. 6. Tokyo: Iwanami Shoten.

Ouchi, William G. 1984. The M-form society. New York: Addison-Wesley.

Pascale, Richard Tanner, and Anthony G. Athos. 1981. The art of Japanese management. New York: Simon and Schuster.

Passin, Herbert. 1967. Society and education in Japan. New York: Teachers College Press, Columbia University.

Patrick, Hugh, and Henry Rosovsky, eds. 1976. Asia's new giant: How the Japanese economy works. Washington, D.C.: The Brookings Institute.

Peterson, Richard B., and Lane Tracy. 1988. "Lessons from labor-management cooperation." California Management Review (Fall): 40–53.

Pfeffer, Jeffrey. 1987. "A resource dependence perspective on intercorporate relations." In Intercorporate relations: The structural analysis of business, edited by Mark Mizruchi and Michael Schwartz, 25–55. New York: Cambridge University Press.

Powell, Walter W. 1990. "Neither market nor hierarchy." Research in Organizational Behavior 12:295–336.

Press Journal. 1985–1996. Nihon handotai nenkan (Japan semiconductor yearbook). Tokyo: Press Journal.

———. 1991. VLSI report: Special survey XI. Tokyo: Press Journal.

Riggs, Henry E. 1983. Managing high-technology companies. Belmont, Calf.: Wadsworth.

Sabel, Charles F. 1994. "Learning by monitoring: The institutions of economic development." In Handbook of economic sociology, edited by Neil J. Smelser and Richard Swedberg, 137–65. Princeton, N.J.: Princeton University Press.

Sakamoto, Yuzaburo. 1990. Hitachi ni miru handotai kojo no genba keiei (Management in Hitachi semiconductor factory). Tokyo: Nikkan Kogyo Shinbunsha.

Sako, Mari. 1992. Prices, quality, and trust: Inter-firm relations in Britain and Japan. Cambridge: Cambridge University Press.

Sangyo Times. 1985–1991. Handotai sangyo keikaku soran (Comprehensive list of semiconductor industry planning). Tokyo: Sangyo Times.

Sato, Ryuzo. 1985. Gijutsu no keizaigaku (Economics of Technology). Tokyo: PHP Kenkyusho.

Scherer, Frederic. M. 1986. Innovation and growth: Schumpeterian perspective. Cambridge, Mass.: MIT Press.

Shepherd, William G. 1997. The economics of industrial organization. New Jersey: Prentice-Hall.

Shimokawa, Koichi. 1990. "90 nendai, sekai saihensei no rikigaku (Mechanism of international restructuring in the 90s." Ekonomisuto (February) 13: 46–49.

Shimura, Takeo. 1984. IC sangyo no shintenkai (New developments in IC industry). Tokyo: Diamond.

Shirosaka, Shunkichi. 1984. Kagaku gijutsushi (History of scientific technology). Tokyo: Nikkan Kogyo Shinbun.

Sjostrand, Sven-Erik, ed. 1993. "On institutional thought in the social and economic sciences." In Institutional change: Theory and empirical findings, 3–31. Armonk, N.Y.: M.E. Sharpe.

Smitka, Michael J. 1991. Competitive ties: Subcontracting in the Japanese automotive industry. New York: Columbia University Press.

Stowsky, Jay S. 1989. "Weak links, strong bonds: U.S.-Japanese competition in semi-conductor production equipment." In Politics and productivity, edited by Chalmers Johnson, Laura D'Andrea Tyson, and John Zysman, 241–74. New York: Ballinger Publishing.

Takahashi, Kamekichi. 1969. *Nihon kindai keizai no ikusei* (Development of Japanese modern economy). Tokyo: Jijitsushin.

Tjosvold, Dean. 1984. "Cooperation theory and organizations." Human Relations 37 (9): 743–767.

Tokyo Daigaku Shakai Kagaku Kenkyusho. 1992. *Gendai Nihon shakai* (Contemporary Japanese society). Tokyo: University of Tokyo Press.

Toyo Keizai, ed. 1982. *IC kakumei kageno shuyakutachi* (Main shadow players in IC revolution). Tokyo: Toyo Keizai Shinposha.

Tsuda, Masumi. 1984. *Nihonteki keiei no daiza* (Base of Japanese management). Tokyo: Chuo Keizaisha.

Ueda, Sojiro. 1978. *Gendai shihon shugi to chusho kigyo keiei* (Modern capitalism and the management of small- and medium-sized firms). Tokyo: Shin Hyoron.

Uekusa, Masu. 1987. "Industrial organization: The 1970s to the present." In The political economy of Japan, Vol. 1, edited by Kozo Yamamura and Yasukichi Yasuba, 465–515. Stanford: Stanford University Press.

United Nations. 1979. Demographic yearbook: Special issue. N.Y.: United Nations.

United Nations Center on Transnational Corporations. 1986. Transnational corporations in the international semiconductor industry. New York: United Nations.

White, Merry. 1988. The Japanese educational challenge: A commitment to children. Tokyo: Kodansha International.

Whitley, Richard. 1992. Business systems in East Asia: Firms, markets and societies. London: Sage Publications.

Williamson, Oliver. 1975. Markets and hierarchies: Analysis and antitrust implications. New York: The Free Press.

———. 1981. "The Economics of organization: The transaction cost approach." American Journal of Sociology 87(3): 548–577.

———. 1985. The economic institution of capitalism. New York: The Free Press.

Yano Research Institute. 1984. The Japanese semiconductor and IC industry. Tokyo: Yano Research Institute.

Yoshihara, Kunio. 1979. Japanese development: A short introduction. Tokyo: Oxford University Press.

Yoshino, M. Y. and Thomas B. Lifson. 1986. The invisible link. Cambridge, Mass.: MIT Press.

Yoshitomi, Mamoru. 1990. "*Keiretsu*: An insider's guide to Japan's conglomerates." Economic Insights (September/October): 10–14.

Young, Allyn. 1928. "Increasing returns and economic progress." Economic Journal 38: 523–542.

Zajac, Edward J. and Cyrus P. Olsen. 1993. "From transaction cost to transactional value analysis: Implications for the study of interorganizational strategies." Journal of Management Studies 30 (1): 131–145.

Subject Index

Author Index